Hello World

To John

ALICE RAWSTHORN

Hello World

WHERE DESIGN MEETS LIFE

HAMISH HAMILTON
an imprint of
PENGUIN BOOKS

Hamish Hamilton

Published by the Penguin Group
Penguin Books Ltd, 80 Strand, London WC2R 0RL, England
Penguin Group (USA) Inc., 375 Hudson Street, New York, New York 10014, USA
Penguin Group (Canada), 90 Eglinton Avenue East, Suite 700, Toronto, Ontario, Canada M4P 2Y3
(a division of Pearson Penguin Canada Inc.)
Penguin Ireland, 25 St Stephen's Green, Dublin 2, Ireland (a division of Penguin Books Ltd)
Penguin Group (Australia), 707 Collins Street, Melbourne, Victoria 3008, Australia
(a division of Pearson Australia Group Pty Ltd)
Penguin Books India Pvt Ltd, 11 Community Centre,
Panchsheel Park, New Delhi – 110 017, India
Penguin Group (NZ), 67 Apollo Drive, Rosedale, Auckland 0632, New Zealand
(a division of Pearson New Zealand Ltd)
Penguin Books (South Africa) (Pty) Ltd, Block D, Rosebank Office Park,
181 Jan Smuts Avenue, Parktown North, Gauteng 2193, South Africa

Penguin Books Ltd, Registered Offices: 80 Strand, London WC2R 0RL, England

www.penguin.com

First published 2013
001

Set in Arial and Helvetica
Typeset by IBO
Printed in Great Britain by Clays Ltd, St Ives plc

A CIP catalogue record for this book is available from the British Library

ISBN: 978–0–241–14530–2

www.greenpenguin.co.uk

Penguin Books is committed to a sustainable
future for our business, our readers and our planet.
This book is made from Forest Stewardship
Council™ certified paper.

ALWAYS LEARNING PEARSON

Contents

Prologue
Hello World

Making the world's resources serve one hundred per cent of an exploding population can only be accomplished by a boldly accelerated design revolution to increase the present performance per unit of invested resources. This is a task for radical technical innovators, not political voodoo-men.
— Richard Buckminster Fuller[1]

Try to imagine that you have never seen a smartphone before. It does not matter what make or model it is; if you came across one for the first time, how could you possibly guess what it was intended to do? Looking at it would not help, because you would not find the slightest hint in that tiny sliver of glass, metal and plastic as to what it is for, or how you might use it.

Would you expect something slimmer than a cigarette packet and not much taller or wider to be more powerful than a bulldozer? No. And would there be anything in its appearance to suggest that it could fulfil the functions of a telephone, camera, Internet browser, digital home management system, games console, DVD player, sound system, watch, clock, diary, address book, barometer, calculator, satellite navigation system, compass and countless other objects too? No again. To anyone who was not already familiar with it, a tiny digital device like a smartphone would seem utterly inscrutable.

The reason why so much computing power can be compressed into so little space is because generations of scientists strove first to invent the transistor, as a means of storing data, and then to make it smaller. So successful were they that several million transistors can now be squeezed on to a tiny microchip, which would have contained no more than a handful when it was introduced in the late 1950s. That is why even the smallest smartphone packs more processing power than an enormous bulldozer engine. It is a remarkable achievement, which demanded courage, vision, knowledge and skill from the scientists involved, but their work may never have left the laboratory and transformed the lives of millions of people, if not for design.

When people talk about the design of a phone, or of most other objects, they are usually referring to how they look, but that is only a fraction of what it will have contributed. It is thanks to the process of analysis, visualization, planning and execution, which we call design, that breakthroughs in scientific research are translated into products, like smartphones, that can make our lives easier and more enjoyable. And it is thanks to the designers who developed the software with which we operate those devices that we can instruct them to perform so many different tasks: waking us up on time; scouring the Internet for information; sending messages to people thousands of miles away; playing music or movies; unlocking the doors of cars or hotel rooms; or, possibly, replenishing the contents of the fridge when we are miles away from home.

If the phone's operating software has been designed intelligently, you should be able to effect all those things effortlessly, because you will know instinctively what to do and when to do it, without ever having to dwell on it. The visual clues that appear on the screen will tell you which symbols to touch, or keys to stroke. But if its design has been botched, the process will be flawed. At best, you will not be able to operate the phone without a struggle. At worst, it will refuse to function, at least not in the way you want it to. In any event, you will feel grumpy, frustrated, possibly furious, and may very well end up blaming yourself for being incompetent, even if it is the designer's fault that you cannot use the phone properly, not yours.

Then there are the possible consequences of its design deficiencies. You may find yourself waking up late because you could not work out how to set the alarm correctly. Missing an important meeting after failing to retrieve a voicemail on time. Stranding an elderly relative in a dangerous place having missed an email. Or hauling heavy luggage up to a hotel room only to discover that your phone cannot unlock the door after all. Those are just a few of the problems and mishaps that could have been avoided if only the phone had been more thoughtfully designed.

Design can empower or disempower us in every other aspect of our lives. Can you find the information you need from a website quickly and easily, or are you baffled by it? Has a signage system guided you deftly and safely from place to place, or left you feeling bewildered? Are you comfortable with the ethical and environmental implications of the things you buy, or anxious about the consequences? Do those new shoes make you feel strong, confident and gorgeous, or silly and unstable with aching ligaments and swollen ankles? And when you last bought a car, did you choose it because you loved it, or because it was the model you disliked the least? Not that it seems fair to complain about uncomfortable shoes or befuddling phones when so many people are struggling to get by on so little, but that litany of problems illustrates design's power to influence our lives – for better and for worse.

When design is deployed wisely, it can bring us pleasure, choice, strength, beauty, comfort, decency, sensitivity, compassion, integrity, ambition, security, prosperity, diversity, camaraderie and so much more. But if its power is abused, the outcome can be wasteful, confusing, humiliating, scary, enraging, even dangerous. And none of us can avoid being affected by design, because it is a ubiquitous element of our world that can determine how we feel, what we do and how we look, often without our noticing. We are able to decide whether or not to engage with many other aspects of daily life: art, literature, theatre, cinema, fashion, sport and music. To some of us, they are irresistible sources of pleasure, while others may find them no more than mildly interesting, or irredeemably dull. Thankfully, we are free to choose the degree to which we will be exposed to each of them. A lot? A little? Not at all? But we cannot escape design, however much we might wish to. All we can do is to try to determine whether its impact on us will be positive or negative, and to do so successfully, we need to understand it, the more thoroughly the better.

This book explores design's influence on our lives now, and in the future. It is called *Hello World* partly as a reference to the first words that flash on to a computer screen whenever a new programming language has been designed successfully. Once a programmer has finished work on a language, like the one that determines how you use your phone, it is customary to test it by running a basic program, on the grounds that if that does not work, nothing will. Most programmers use the same test program, one developed at the turn of the 1970s by Dennis M. Ritchie, a research scientist at Bell Laboratories in New Jersey, where the transistor was invented.[2] All it does is spell out 'hello, world'. Year after year, thousands of programmers have waited for those words to appear to reassure them that their efforts have been successful. But there is another reason for choosing *Hello World* as the title of this book, because it is such an eloquent way of describing what design does, and how it affects us.

Design is a complex, often elusive phenomenon that has changed dramatically over time by adopting different guises, meanings and objectives in different contexts, but its elemental role is to act as an agent of change, which can help us to make sense of what is happening around us, and to turn it to our advantage. Every design exercise sets out to change something, whether its intention is to transform the lives of millions of people, or to make a marginal difference to one, and it does so systematically. At its best, design can ensure that changes of any type – whether they are scientific, technological, cultural, political, economic, social, environmental or behavioural – are introduced to the world in ways that are positive and empowering, rather than inhibiting or destructive.

The process of design existed long before a word was invented to describe it. Whenever human beings sought to change their way of life or their surroundings, by barricading a prehistoric cave against predators or modifying the head of a spear, they acted as designers, but did so instinctively. All that changed in the industrial age, when design was redefined as a commercial tool. Design achieved great things in that guise by helping to make billions of people safer, healthier, more enlightened, happier and kinder. Information designers guided us around the world, and communication designers explained it to us. Aerospace designers gave us the confidence that if we strapped ourselves into the seats of oddly shaped metal contraptions, we would be flown thousands of miles over oceans and mountains to be deposited on the ground, more or less where we wished to be. Fashion designers amused us, though not always intentionally. And it is thanks to software and product designers that we engage on a daily basis with technologies which were the stuff of science fiction only a few years ago.

But design has also been trivialized, misunderstood and misused. It is routinely confused with styling, and with expensive, uncomfortable chairs or silly shoes with vertiginously high heels. It is typecast as a seductive ploy that tricks us into buying things of questionable value, which we will swiftly tire of and abandon with the rest of the toxic junk dumped in bloated landfill sites. All too often, it is seen as an indulgence for spoilt consumers in developed economies, rather than as a means of helping the disadvantaged escape from poverty.

Hello World will explore whether design can defuse those misconceptions to make a deeper, more meaningful contribution to our lives in future, at a time when we face colossal changes on so many fronts, unprecedented in their speed, scale and intensity. Advances in science and technology are accelerating. The environmental crisis is deepening, as is the economic gulf between rich and poor. The social and political systems that regulated society in the last century are breaking down. The logistics of our daily lives and the objects that fill them are dramatically different from ten years ago, and will be as different again in ten years' time, making a nonsense of many of the old assumptions about how we wish to live and what we value.

All these changes present daunting challenges for design. How can it help us to make the most of future breakthroughs in nanotechnology, supercomputing, biomimicry, supramolecular chemistry and the findings of the physicists working at the Large Hadron Collider in Switzerland and on the new super-telescopes in northern Chile? Develop new forms of transport, which are less stressful, dangerous, expensive and destructive? Ensure that we have no reason to feel guilty about any aspect of how something we buy is designed, made, tested, shipped, sold and will, eventually, be disposed of? Make productive use of the tsunami of data surging out of ever more powerful computers? Do the same for three-dimensional printing and other digital production technologies? Clean up the tens of millions of pieces of space junk hurtling around outer space? Reinvent dysfunctional social services? Stop computer viruses? Redress the tragic imbalance of resources that leaves so many people with too little, and the rest of us with more than we need or want?

Design is not a panacea for those problems, far from it, but it is one of the most powerful tools at our disposal to resolve them, providing it is deployed intelligently. The more we know about design, the stronger our chances will be of using its power wisely, and the first step is to get to grips with what it is.

1 What is design?

Designing is not a profession but an attitude.
— László Moholy-Nagy[1]

When it came to waging war, outwitting enemies, crushing dissent and terrorizing foes, Ying Zheng had all the necessary skills. He may not have been quite as bloodthirsty and remorseless a tyrant as his harsher critics have claimed, but he was exceptionally cunning, determined and resourceful. And these qualities proved invaluable in his efforts to transform the obscure kingdom of Qin into one of the most powerful and enduring empires in history: China.

Ying Zheng was in his early teens when he became king of Qin in 246 BC.[2] It took nearly ten years for him to foil his political rivals and assert his full authority over the realm. A decade and a half later, he had conquered all of Qin's wealthier, seemingly more powerful neighbours. The military coups began when his forces attacked and defeated the weakest of the neighbouring states. Another conquest was made by pretending to flee, and ambushing the opposing army when it pursued them. A third state surrendered after Ying Zheng's troops flooded its capital city by diverting the Yellow River. In the final battle, the ruler of Qin admitted defeat after Ying Zheng's forces staged a surprise attack on his capital, arriving there before its defences were prepared.

Ruthlessness, military might, tactical prowess and diplomatic guile served Ying Zheng well, as did an expertly managed espionage network of informants, a redoubtable arsenal of weapons and a rigorously trained army of archers, swordsmen, charioteers and cavalrymen. But he had another asset at his disposal, one that is often overlooked by military historians but has proved indispensable to ambitious warlords throughout history: an unusually sophisticated understanding of design.

There were many decisive elements in Ying Zheng's military triumphs, but an important one was the design of his troops' weapons. When he embarked on his career as a warlord, an army's fortunes were determined by the combined might of hundreds of thousands of foot soldiers, mostly peasants who had been conscripted, often against their will. Training these *ingénu* soldiers and ensuring that they could use their weapons efficiently was essential to the success of any military campaign.

The design of all weaponry was improved under Ying Zheng's command. The optimum size, shape, choice of material and method of production for each piece was determined, and every effort made to ensure that weapons of the same type adhered to the chosen formula. The Qin army had used bronze spears for over a thousand years, but the blades were rendered shorter and broader. The dagger-axes were redesigned too. Putting six holes in the blades, rather than four, ensured that their bronze heads could be attached more securely and were less likely to shake loose in the frenzy of battle.

Even more important were the changes to Qin's bows and arrows. Archers were critical in determining the outcome of every stage of combat in Ying Zheng's era, but their weapons were made by hand, often to different specifications. If an archer ran out of arrows during a battle, it was generally impossible for him to fire another warrior's arrows from his bow. Similarly, if he was killed or injured, his remaining ammunition would be useless to his comrades. And if a bow broke, that archer's arrows risked being wasted. The same problems applied to more complex weapons like crossbows. The result was that an army's progress was often impeded by weapons failure because its archers were unable to fight at full efficiency, if at all.

Ying Zheng's forces resolved these problems by standardizing the design of their bows and arrows. The shaft of each arrow had to be a precise length, and the head to be formed in a triangular prism, always of the same size and shape. The components of longbows and crossbows were made identical too, and these design formulas were rigidly enforced. Each piece of government equipment was branded with a distinctive mark to identify who had made it and in which workshop. If a particular weapon was deemed substandard, the offending artisans would be fined, and punished more severely if the problem recurred. To foster a sense of collective responsibility for the workshop's output, all their co-workers and supervisors were punished too.[3] Unsurprisingly, the efficacy of Ying Zheng's arsenal increased significantly, with disastrous consequences for his less resourceful foes. Centuries later, there are echoes of the rigour with which he and his henchmen improved the design of their weapons in the design studios of companies like Apple and Samsung, where teams of designers strive to find cheaper and more effective ways of producing new digital devices that are lighter, sleeker, faster and more versatile than their predecessors.

Having established his empire and renamed himself Qin Shihuangdi, or 'First Emperor of China', in 221 BC, he imposed his authority on other aspects of his subjects' lives. A unified system of coinage was introduced, as were standardized weights and measures, a universal legal code and common method of writing.[4] These changes made daily life more orderly, and boosted the economy by making it easier for people from different regions to trade. They also had a symbolic importance in helping to persuade the new emperor's subjects, many of whom had fought against his army in battle, or had family or friends who had died doing so, that they had a personal stake in his immense domain. Take the new coins. Every time a farmer or a carpenter used them, they saw a tangible reminder that they themselves were part of a dynamic new empire, and had good reason to feel grateful to its visionary founder and ruler.

To reinforce the new national identity, Qin Shihuangdi forced the wealthiest and most powerful of his subjects to leave their ancestral homes, thereby severing their ties to the local communities, and to settle in Qin's capital, Xianyang. Hundreds of palaces were constructed there, including sumptuous ones for the emperor himself, which were positioned to align with the stars in an allusion to his role as the ruler, not only of China, but the rest of the universe too.[5] Such trappings have been the default means with which egomaniacal rulers have flaunted their power throughout history, hoping to dazzle and intimidate their subjects, but Qin Shihuangdi's were especially magnificent. He also made sure that the inhabitants of even the most remote regions knew of his power and achievements by ordering descriptions of his feats to be carved into mountains across China.

Here, Qin Shihuangdi demonstrated his command of design strategy, by applying what we now call design thinking to identify what he needed to do to secure the future of his regime, and to communicate the results to his subjects. There are parallels between his strategic use of design and its role in successful corporate identity programmes, such as Nike's, and communication exercises like Barack Obama's presidential election campaigns. Yet the first emperor of China also applied design as a medium of self-expression, and it is for this that he is most famous.

Like his immediate ancestors, Qin Shihuangdi believed that the living and the dead belonged to the same community, and regarded death with dread, as an attack on the individual by evil spirits. He tried desperately to avoid it, dispatching expeditions into the seas east of China in doomed attempts to find the mythical islands of Penglai, Yingzhou and Fangzhang, whose herbs and plants were thought to have the power to grant everlasting life. When the expeditions failed, the emperor decided to ensure that, if he had to die, he would, at least, enjoy a spectacularly self-indulgent afterlife. All his ancestors had built lavish tombs, but Qin Shihuangdi was determined that his resting place would be even more magnificent, reflecting his greater power and status. He ordered the construction of an imposing burial chamber in a vast underground palace covering an area of over twenty square miles on the slopes of Mount Li, near what is now the Lintong area of China, where a group of local farmers discovered its remains by chance in 1974, while drilling for water.[6]

Like his forebears, Qin Shihuangdi would have insisted that the same courtiers, servants, warriors and entertainers who had served him during his life should continue to care for him in death. Some members of his retinue would have been expected to die with him, like his favourite pets and horses and the finest animals in his menageries, but most of his posthumous entourage were replicas of the originals made from bronze, wood or terracotta. Many of them were intended to entertain the dead emperor, but he was particularly anxious to guarantee his personal safety in the afterlife, fearful that the victims of his warmongering would attempt to wreak revenge upon him in death. One set of pits to the west of his burial chamber contained a cavalry of elaborately crafted bronze horses and chariots, each of which was half life-size and made from over two tonnes of bronze, embellished with gold and silver ornaments. To the east of the chamber were three pits whose contents have made his tomb one of the most famous archaeological sites in history: the 'Terracotta Army' of seven thousand life-sized warriors who were to guard the dead emperor, armed with real swords, daggers, crossbows and axes.

Hundreds of thousands of workers were transported to Mount Li to construct the imperial tomb and its contents. A dedicated team of over a thousand labourers toiled on the terracotta warriors, whose production demanded an even more sophisticated system of standardized design than Qin's military arsenal had done. Each figure was made to particular specifications by coiling, rolling and moulding the terracotta, and applying the decorative details by hand. The work was so arduous, and took so long, that many of the workers died on site and were buried there.[7]

Just as Qin Shihuangdi had deployed design with extreme efficiency to amass wealth and power during his life, he used it to secure what he believed would be an equally resplendent death, by creating the afterlife of his fantasies, which served a practical purpose too. Building such an outlandishly extravagant burial site was so eloquent a testimony of his might that it reinforced it as effectively as his celestially planned palaces, mountain inscriptions and the new imperial currency. But it was also a physical manifestation of the inner world of his imagination, a material expression of how China's first emperor saw himself, and wished to define his place in history, which presaged contemporary design spectacles such as Olympic Games opening ceremonies, the Arirang Festivals in North Korea and the elaborate sets of Chanel's haute couture shows at the Grand Palais in Paris.

What would design have meant to Qin Shihuangdi and his subjects in the third century BC? Probably nothing, at least in its contemporary sense. The word 'design' dates back to ancient Rome and the Latin verb *designare*, which had several meanings, including to mark, trace, describe, plan and perpetrate. But it is highly unlikely that Qin Shihuangdi would have been congratulated on his design prowess, even though he deployed different interpretations of the sequence of thought and actions, which we describe as design, with exceptional rigour and ingenuity. And at every stage of his career, first as a preternaturally ambitious young warlord, and then as one of the richest, most powerful rulers in history, design played a decisive role in his success. Yet unlike latter-day design tacticians such as Apple, Chanel, Nike, Barack Obama's campaign advisors and the despotic Kim dynasty, Qin Shihuangdi conceived and executed his design feats entirely instinctively.

So did everyone else who, equally unconsciously, engaged with design long before and after his reign. As the American industrial designer Henry Dreyfuss wrote in his 1955 book *Designing for People*: 'Somewhere in the deep shadowy past, primitive man, desiring water instinctively dipped his cupped hands into a pool and drank. Some of the water leaked through his fingers. In time he furnished a bowl from soft clay, let it harden and drank from it, attached a handle and made a cup, pinched the rim at one point, creating a pitcher.'[8] When Dreyfuss's thirsty caveman set about transforming that bowl into a cup, and then a jug, he acted on intuition, because he needed to find a way of drinking without wasting water, just as the young Ying Zheng recognized the need to make deadlier weapons. Whenever medieval carpenters or blacksmiths took it upon themselves to do more than copy an existing object, maybe by making something less fragile or more elegant, they too were acting as de facto designers, as were all the sailors who have ever tinkered with the rigging of their ships, or farmers with their tools and the walls they have built to protect their land.

The first definition of 'design' in the *Oxford English Dictionary* dates from 1548, when it appeared as a verb meaning to 'indicate' or 'designate'. Other interpretations soon emerged. In 1588, it was used as a noun meaning 'purpose, aim, intention'. Five years later, it had assumed a more elaborate role as: 'a plan or scheme conceived in the mind of something to be done, the preliminary conception of an idea that is to be carried into effect by action'.[9] These early meanings have survived, and new ones have appeared over the years, some of them generic, and others defining the word more precisely.

Every word with a long history is redefined over time, reflecting anything from the prevailing attitudes of particular eras and commercial opportunism to unexpected catastrophes, but few words have ended up being as ambiguous as 'design'. The more meanings it has acquired, the slipperier and more elusive it has become, not least because so many of its newer interpretations sit oddly with older ones. The design historian John Heskett has compared the difficulty, if not impossibility, of defining 'design' to doing the same for 'love'.[10] Both words have so many layers of meaning that they can be read very differently in different contexts. Just as 'love' can describe anything from tender affection and lifelong devotion to unbridled lust and destructive obsession, it is possible for 'design' to convey a minute technical detail to one person; a million-dollar chair to another; or a life-changing innovation, such as an efficient, inexpensive prosthetic limb, to a third. As for the members of the neo-creationist movement, who believe that parts of the universe were created, not by the organic process of evolution described by Charles Darwin, but by a rational system orchestrated by an 'intelligent designer', to them design is an explanation for the origins of life.[11]

For as long as there is no clear consensus on what design

Ancient design ingenuity, the stones of Hagar Qim Temple in Malta

A limestone wall on Inis Meáin
in Galway Bay, Ireland

A drystone wall in Chilmark on
Martha's Vineyard, Massachusetts

A drystone wall on the
Shetland Islands, Scotland

represents, it will be prey to more muddles and clichés. The seeds of the problem lie in the sixteenth century, when it ceased to be a purely generic word and was adopted as a specific term by architects, engineers, shipbuilders and artisans to describe the drawings, plans or diagrams which indicated in detail how their work should be built. When Giorgio Vasari visited Leonardo da Vinci's studio in Florence to research his 1550 book *Lives of the Artists*, he recorded seeing his 'designs for mills, fulling machines and engines', as well as 'a splendid engraving of one of these fine and intricate designs with the words in the centre: Leonardus Vinci Academia'.[12]

The objective of such 'designs' was to provide guidance for the craftsmen who would fabricate their contents, with the aim of ensuring that the end result was as the architect or designer had envisaged. This role seems clear enough, but design was soon also used to describe the thought process behind the development of the ideas represented in those 'designs', as well as their outcome. John Heskett summed up this etymological tangle with the sentence: 'Design is to design a design to produce a design.'[13] Silly though it sounds, that phrase is grammatically correct, and factually accurate when used in the right context. And as if that was not confusing enough, the technical definition of 'design' was to be redefined again in the industrial era.

By the early seventeenth century, a delegated system of design and production akin to the one used by Leonardo and his peers was adopted by the porcelain makers in the Jingdezhen area of China, who had established a thriving trade in exporting oriental porcelain to Europe.[14] Decades later, a similar process was introduced to the royal manufactories founded in France by Louis XIV, the *roi du soleil* or 'Sun King', to produce exquisitely crafted furnishings for his palaces. The most famous of them was Gobelins, the Parisian tapestry and cabinetmakers, where the artist Charles Le Brun cast himself in the unofficial role of chief designer.[15]

A personal favourite of both Louis and Jean-Baptiste Colbert, his powerful finance minister, Le Brun eventually became a director of Gobelins and saw his work there as a way of extending his aesthetic influence over the royal court. Hundreds of craftsmen were employed in Gobelins's workshops, where they translated detailed drawings and models provided by Le Brun and other artists into finished products. To the king and Colbert, the magnificent workmanship of the manufactories formed part of their strategy of convincing the French people that Louis XIV was the world's greatest living monarch and a figure of inestimable historic importance: exactly the same impression that Qin Shihuangdi had sought to make on his subjects centuries before. The awe-inspiring effect of Louis's furniture was reinforced by the splendour of his palaces, gardens, horses, weaponry, wardrobe, the statues of him erected throughout France, and every other visual manifestation of himself and his reign.[16]

Impressive though the king and Colbert were in their strategic use of design, its practical role in their manufactories proved to be more influential. When the Industrial Revolution began in the late 1700s, the Gobelins system was implemented on a larger scale by newly opened factories, which needed to find efficient ways of manufacturing their goods to the same specifications at consistent quality in huge quantities, just as Qin's military arsenal had done. In the forefront was the British industrialist Josiah Wedgwood, who personally supervised the development of the products made in his Staffordshire potteries.[17] His success prompted other manufacturers to copy his methods, which became a template for what we now recognize as the industrial design process.

Industrialization defined design for the modern age by categorizing it into different disciplines and job descriptions, as well as by spawning a cottage industry of design strategists and consultants. By classifying design, it created the illusion of clarity, yet design is no less confusing in its industrial context than in any other. How can its use in the development of something as huge and as mechanically complex as a train compare to that of a Wedgwood vase? It cannot. Nor can it be compared to the design of the other components of rail travel: the tracks, bridges, tunnels, stations, signs, maps, timetables, tickets, signal boxes, signalling systems, the railway company's visual identity, staff uniforms and so on. Or to other fields, like health, education, warfare, fashion, leisure, urbanism, flight, cars, computing and space exploration. And consider the very different dynamics of the working practices of various design disciplines: from an haute couture fashion designer, whose tools may be paper, a pencil, pins, scissors, fabric and trimmings; to the inscrutably complex supercomputers and wind tunnel tests in the laboratorial workspaces of aerospace design teams.

Precisely what design means and how it is applied has differed from company to company, industry to industry, country to country and era to era, depending on the scale of the project, its technical complexity, legislative constraints, political sensitivities, and random factors such as whether the engineers tend to shout louder than the designers in meetings. What is referred to as 'design' in one situation may be called 'styling' in another, or 'engineering', 'programming', 'art direction' or 'corporate strategy'. And outside the commercial sphere, any intuitive design exercise is often ascribed to resourcefulness or common sense. Not that this should be surprising. Design's versatility and its ability to adapt to so many different contexts are among its greatest strengths. But the lack of coherence in its industrial role has fostered more misunderstandings.

One is to relegate design to the role of styling or decoration. Few things infuriate designers more, and 'style' has become a dirty word for design purists, yet when the word 'design' features in the media or advertising campaigns, it is often synonymous with style. To some degree this is inevitable, not least because styling is the area of design on which people tend to feel most confident about expressing an opinion, especially as the logistics of the industrial design process can seem so complex and technologically intensive that it appears dauntingly opaque. Few of Josiah Wedgwood's late eighteenth-century customers could have analysed the quality of the clay used to make a plate, but they would have known instinctively whether or not they warmed to its floral pattern and particular shade of blue – just as most of us feel more confident in saying whether we like the look of a new computer or digital device than analysing if its operating software is up to scratch, or if the manufacturer truly has done everything possible to ensure that its far-flung subcontractors are treating their employees fairly. Besides, styling is the one aspect of industrial design that no one else bothers to claim. Design can also contribute to corporate strategy, engineering, sustainability, communications, branding, production planning, sourcing and social responsibility, but so do other areas of the organization, such as the finance and marketing teams, which are often loath to share the credit.

Another misconception is to regard design solely as a commercial tool. In the early 1900s, the Constructivist movement hailed industrial design as a means of building a better world by producing 'a new thing for the new life' in post-revolutionary Russia.[18] It was seen in the same heroic spirit by the modernist pioneers in the De Stijl group in the Netherlands and at the Bauhaus school in Germany. But by the second half of the twentieth century, design was generally consigned to a commercial role, not least because it played it so well: enriching entire economies as well as individual businesses. In post-war Italy, a new generation of industrialists joined forces with gifted young designers, most of whom had trained either as architects, like the Castiglioni brothers, Carlo Mollino and Ettore Sottsass, or as artists, in the case of Joe Colombo, Enzo Mari and Bruno Munari. Together, they produced such appealing and ingenious products that consumers were persuaded to pay extra for 'the Italian line'.[19] The seductive image of *La Dolce Vita* Italy sold millions of Fiat cars, Vespa scooters, espresso machines, Castiglioni lights for Flos, and Sottsass typewriters for Olivetti. As well as powering Italy's post-war recovery, the 'design adds value' formula has been a template for expanding economies ever since, from Japan in the 1960s to China and South Korea today. As IBM's former president Thomas J. Watson Jnr put it in a lecture he gave at the Wharton School of Business in 1973: 'good design is good business'.[20]

Even the intellectual discourse on design often focused on its commercial guise. One of the most influential design commentators of the late twentieth century was the British cultural historian Reyner Banham, who produced a remarkable series of texts analysing its relationship to consumer culture and the fledgling pop movement.[21] His take on design was reflected in the writing of the French philosophers Roland Barthes and Jean Baudrillard, whose books unpacked the underlying meaning and symbolism of consumer products,[22] as did the artists whose work proved most incisive in interrogating design, such as Sigmar Polke and Hans Haacke in Germany, Ed Ruscha in the United States, and Richard Hamilton in Britain.

There have always been alternative approaches to design, which have spurned its commercial applications, in favour of deploying it to improve our quality of life. In the same year that Watson Jnr delivered his Wharton lecture, a group of designers and programmers at the Massachusetts Institute of Technology set up the Visual Language Workshop (VLW), whose research has had an enormous influence on the digital images we now see on our phone and computer screens. Among them was Muriel Cooper, a graphic designer, then in her late forties, who was renowned for creating eloquent covers for books, including Robert Venturi, Denise Scott Brown and Steven Izenour's 1972 postmodernist text, *Learning from Las Vegas*.[23]

Cooper discovered computers by chance in 1967, when

she wandered into a class taught by her MIT colleague Nicholas Negroponte, and was flummoxed by the coded data she saw on the screen. 'Didn't make any goddamned sense to me,' she recalled.[24] Baffled though she was, the experience convinced her of the need for digital imagery to be as expressive and appealing as possible, like the best of printed graphics. At the time, computers were the preserve of programmers, like Ron MacNeil, with whom she co-founded the VLW. Helped by him, Cooper insisted on applying design solutions to what was then seen as a technology problem. A succession of gifted software designers studied under her at the VLW, where she worked until her death in 1994 with her black poodle, Suki, snoozing in her office. Among them was John Maeda, who recalled her habit of hoisting her feet up on to the desk as a tactical ploy whenever powerful men challenged her. 'In Muriel's era, men were tough, and she said: "I'll be tougher." So she showed them.'[25]

Watson Jnr's lecture also coincided with the end of the 'World Design Science Decade' launched by another American design maverick, Richard Buckminster Fuller. Universally known as 'Bucky', he had championed environmentally responsible design since the 1920s, in famously long, often incomprehensible lectures, which also touched on his pet theories of universal patterns, the marvels of the tetrahedron and the nutritional value of Jell-O. One lecture ran for forty-two hours and was entitled 'Everything I Know'.[26] Bucky was so prolific that the *New Yorker* described him in a 1966 profile as 'an engineer, inventor, mathematician, architect, cartographer, philosopher, poet, cosmogonist and comprehensive designer'.[27] Many people in those fields dismissed him as a dotty eccentric, and some of his most cherished schemes flopped, notably a floating city and flying car, but his emergency shelter – the geodesic dome – was among the twentieth century's most successful examples of humanitarian design. Bucky's formula for constructing a make-shift dome on any terrain, even in extreme weather, from scraps of wood, pieces of metal, old clothes or blankets, and whatever else is available, has provided sorely needed shelter for hundreds of thousands of people, often in desperate circumstances, all over the world. Producing 'more for less' was Bucky's overriding objective,[28] and the 'World Science Decade', which ran from 1965 to 1975, was intended to nurture a new generation of 'comprehensive designers' cast in his mould, who would dedicate their working lives to planning a fairer, more productive future for mankind.[29]

Neither Buckminster Fuller nor Muriel Cooper had squads

of PR spinners to publicize their achievements, nor did the anonymous designers of the continuing stream of intuitive design coups, unlike all those who have stood to profit from reminding us of design's commercial accomplishments. Design consultancies hoping to win new clients by proving how effective they have been for existing ones. Corporate executives whose best bet of clambering up the promotional ladder is to prove how lucrative their past design decisions have been. Politicians seizing the chance to make grandiose claims for economic regeneration policies. Business-school professors drumming up interest in design management courses. No wonder the media coverage of 'design' tends to focus on the nine ambulances and an emergency control vehicle that were called when irate shoppers came to blows over cut-price £45 sofas on the opening day of an IKEA furniture store in north London; or the raw eggs being pelted at the windows of an Apple store in Beijing by angry hordes of people, who had waited outside for hours to buy a new model of the iPhone on its first day on sale, only to be told that it would not open, because Apple feared for the safety of its staff and customers.[30]

Not that they are intended to, but such reports aggravate suspicions that design can be shifty, deceitful, manipulative and untrustworthy. As long ago as 1704, design was cast in a malevolent guise when it was described, according to the *OED*, as 'a crafty contrivance' or 'a scheme formed to the detriment of another'. The dictionary's first definition of a designer includes two very different interpretations of the role: 'one who designs or plans'; and 'in bad sense, a plotter, schemer, intriguer'.[31] If you were told that I had 'designs on you', or on something that you valued, you would be justified in feeling alarmed. The use of the word 'designer' as an adjective has proved equally divisive. If an object is described as a 'designer sofa', 'designer hotel' or 'designer shoe', depending on your perspective, you would either expect it to be so enticing that you would cheerfully pay more for it than a less ostentatious version of the same thing, or you would darkly suspect it of being a rip-off.

Predictably, design has often ended up being misinterpreted or dismissed as a styling ruse. Unfair and clichéd though such stereotypes are, they have been remarkably effective in distorting perceptions of design by obscuring its constructive qualities. The danger is that by doing so, they may be preventing design from being given the chance to prove its worth in other areas of our lives. To draw a parallel with health care, if a hospital decided that its staff would treat physical problems, but not psychological issues, it would provide a sorely inadequate service and its patients would suffer unnecessarily. The same applies to design. If governments, businesses, banks, educationalists and NGOs think of it as only being useful for, say, producing fast cars and photogenic dresses, it risks being restricted to those roles, and is likely to be overlooked when it comes to tackling other challenges. Why would it occur to them that design might make a constructive contribution to the environmental and humanitarian causes championed by Buckminster Fuller, if it is only ever depicted as something that incites hysteria over sofas and phones?

Thankfully, that is changing. Firstly, a new generation of designers has rejected the constraints of commercial design and cast themselves as contemporary incarnations of Bucky's 'comprehensive designers'. Sustainability is high on the agenda, as are humanitarian challenges, such as economic development and improving social services. Take Nathaniel Corum, an American designer who started out in commercial design but wearied of 'shopping for gold-plated fixtures', and has since divided his time between running the education programme of the volunteer network Architecture for Humanity, helping to plan its post-earthquake reconstruction effort in Haiti, and building sustainable, off-grid homes for Navajo elders in New Mexico and Arizona. His commitment to humanitarian causes was inspired by his experience as a student working with Berber communities in Morocco, and tribal groups in Montana and North Dakota. Using AfH's office in San Francisco as his base, Corum leads a nomadic life equipped with a military standard computer, which is shockproof, waterproof and so crushproof that it would still work if a car was driven over it.[32] Similarly, the British social scientist Hilary Cottam worked on urban poverty projects for the World Bank in Africa, before founding the social design group Participle, which is committed to 'addressing the big social issues of our time'. Participle forms teams of designers and other specialists to work with local and national government to plan and deliver more efficient ways of dealing with acute social problems, such as ageing, poverty, crime and intolerance.[33]

Critically, designers like those working for Participle have been given the chance to tackle such endeavours. Partly because the social, ecological, political and economic challenges confronting us are now so grave,[34] and partly because many of the conventional methods of dealing with them have become decreasingly effective, there has been a growing willingness to experiment with new approaches, including design, not least as there has been a fundamental change in design practice.

Traditionally, design was valued chiefly for the things it produced, whether they were tangible, such as objects, spaces and images, or intangible like software. The design process is now deemed to have a value of its own and is increasingly applied to strategic and organizational issues in the form of what is called 'design thinking'. Coined in 1991 by the American design engineer David Kelley, design thinking refers to the skills that designers develop, often without realizing, in analysing problems, using lateral thinking to identify smart solutions, and persuading other people to embrace the outcome.[35] Qin Shihuangdi deployed a process very much like design thinking to assert his authority over his new subjects and to imbue them with a sense of national identity, as did Josiah Wedgwood when planning the expansion of his ceramics business. It is tempting to interpret the vision and confidence that Wedgwood displayed in making gutsy investment decisions – such as building a network of canals to transport his fragile wares by water, rather than packing them into horse-drawn carriages on bumpy eighteenth-century lanes[36] – to the intuitive design skills he nurtured when developing new products for his potteries.

By liberating the design process from its traditional outcome, design thinking has enabled designers to apply their innate strengths to a wider range of challenges. It has been deployed as a commercial tool by companies like IDEO, which was co-founded by Kelley in 1991 in Palo Alto, California, originally to develop technology products for nearby Silicon Valley companies, and now advises banks on how to devise new types of accounts, and health-care groups on improving patient care.[37] Design thinking has also proved indispensable in the public realm, not only to Participle, which uses it in all of its social design projects, but to the graphic designer Peter Saville, in his strategic role for the city of Manchester. Having made his name in the 1980s by designing record sleeves for bands like Joy Division and New Order, Saville has advised the city council on a range of issues, such as proposed property developments and the biennial cultural event the Manchester International Festival, often using the same thought processes with which he once devised Joy Division's artwork.[38]

What does this mean in practice? Consider one urgent

issue: the need to replace a dysfunctional road transport system with a safer, cheaper, cleaner, less stressful alternative. The conventional design solution would be to develop new forms of energy-efficient vehicles, which may very well help to alleviate the environmental damage, but would not solve all of the problems. That will require more radical changes in the design of the roads themselves and every other aspect of traffic management. Encouraging people to share journeys would be one option, as would coaxing them into cycling rather than driving, and trying to spread the flow of traffic through the day, possibly by introducing dynamic pricing for fuel, road tolls and parking spaces. Up until recently, a designer's role in such endeavours would have been restricted to producing leaflets or websites explaining what courses of action economists, politicians, social scientists, psychologists, statisticians and corporate strategists had decided to take, but in future, they could influence decision making, as Participle and Peter Saville have done.

It would be foolish to overstate design thinking's potential, but it has already proved its worth and, if it continues to do so, may be deployed increasingly as an alternative to conventional approaches to design. By removing the technical aspects of the design process, it has also made it easier for people from other fields to participate in design exercises, and for designers to contribute to theirs. And by opening up the design process, design thinking has reasserted the instinctive qualities of resourcefulness and ingenuity that characterized design in the pre-industrial age.

One of the most influential architecture exhibitions of the 1960s was 'Architecture Without Architects', which opened at the Museum of Modern Art, New York, in 1964. Curated by the historian Bernard Rudofsky, it explored the design of buildings by people outside the architectural profession, including sled houses and cliff dwellings.[39] The Chinese artist and political activist Ai Weiwei addressed a parallel phenomenon in 'Un-Named Design', an exhibition at the 2011 Gwangju Design Biennale in South Korea, which featured design projects executed by scientists, hackers, farmers and activists, but not professional designers. Among them was the programming code for a computer virus, a bucket made from a basketball by a Chinese farmer, a metal cage converted into a shelter by a homeless man in Hong Kong, a prosthetic leg, and a plan of action for a political protest in Cairo during the Arab Spring.[40]

With the possible exception of the prosthesis, none of

those projects would conventionally have been considered to have been 'designed', yet elements of the design process are recognizable in the development of each one. Heartening though it is to celebrate the farmer's ingenuity in reinventing the basketball, and the strategic flair required to plan the Cairo protest, 'Un-Named' begs the question of whether there is anything to be gained from identifying such initiatives as 'design'. After all, there is a fine line between design and common sense, and just about any activity which involves originality and forethought can, in theory, be described as design. Surely the only justification for doing so is if it would improve the outcome?

Arguably, it might. Had the anonymous designers in 'Un-Named' been aware that there was a design dimension to their work, would they have devoted more time and energy to it? Possibly by planning more rigorously? Or by pushing themselves harder to think of unexpectedly apt solutions? The most important contributions to the Cairo protest plan would have been the activists' knowledge of the relevant political issues and local geography, and their ability to motivate fellow activists. Even so, the more thoughtfully considered the protest was, the greater its chances of being successful, and of its protagonists avoiding arrest, or worse.

If we apply the same principle to everyday activities, there are impromptu elements of design in many aspects of our lives. Consider food. If you follow a recipe, you cannot be described as having designed the end result, because you will have produced it formulaically without the element of change that is essential to design. But if you improvise by deviating from the recipe or by making it up as you go along, then you could claim to have 'designed' the dish. Will describing that process as 'design' make a positive difference by making the food taste better? Possibly. Other skills, such as your ability to choose the best ingredients, knowledge of their idiosyncrasies and dexterity with cooking tools, will be more important, but the dish may very well seem more tempting had you paid more attention to the design elements of its preparation. Perhaps by being bolder in combining different flavours. Or by taking greater care over its presentation.

Adopting a more fluid definition of design, and recognizing its potential outside a commercial context, will not bring greater clarity. On the contrary, it is likelier to make design seem even more ambiguous by adding multiple new roles to its established ones. Yet, elementally, design will remain the same. Whether its objective is to produce a smaller, more user-friendly smartphone, a loftily ambitious project to reduce carbon emissions, or a makeshift shelter for someone who has lost their home, design will continue to be, as it always has been, an agent of change that can help us to organize our lives to suit our needs and wishes, and wields immense power to influence them, for better and for worse.

Perhaps the most compelling argument for embracing a more eclectic understanding of design is to consider the danger of continuing to ignore its potential. An example is dog breeding, where decades of anarchic, ill-considered attempts to control the development of different breeds of dogs have resulted in a design calamity with tragic consequences. There were no more than fifteen recognized dog breeds in Britain in 1815, but now there are over four hundred, and one hundred and fifty of them have official pedigree societies.[41] Some of those breeds were nurtured for practical reasons. The loyal Dalmatian was originally a guard dog in the mountainous Croatian region of Dalmatia, then a fashionable carriage dog in eighteenth-century England. The plucky Lakeland terrier was bred to forage for vermin in the desolate valleys of the Lake District. The spirited Jack Russell dates back to Trump, a small white and tan terrier whose flair for flushing out foxes so impressed her hunt-loving owner, the Reverend John Russell, that he decided to breed more dogs just like her in early nineteenth-century Oxford.[42] But other types of dog have been bred for their aesthetic attributes, generally ones that endear them to humans and make them more marketable, such as tiny 'teacup' chihuahuas and goofy bulldogs with big heads and baby-like faces. Some of these dogs are, thankfully, robust, but others are sadly prone to hereditary weaknesses having been drawn from such narrow genetic pools for reasons relating to their appearance, rather than to their temperament.[43]

King Charles spaniels are among the most alarming examples, with brains too big for their skulls, while basset hounds are at risk of paranoid delusions, and Dobermann pinschers of narcolepsy.[44] Bulldogs are vulnerable to a litany of health problems, including respiratory illnesses, skin infections, neurological disorders and difficulties with their eyes and ears. The impact on the health of individual dogs has become so grave that some kennel clubs refuse to register puppies with closely related parents, and responsible breeders are trying to create healthier versions of particularly vulnerable breeds.[45] No one set out to redesign a species so radically. And it is difficult to believe that any breeder, not even the unscrupulous puppy-mill owners who cashed in on the craze for baby-faced bulldogs and teeny chihuahuas, intended to cause so much pain by producing sickly dogs with poor health and short life expectancies, yet some of them have done so. The genetic manipulation of the dog is a tragic example of what can happen when a sequence of changes is misconceived and poorly planned, rather than being designed efficiently. Would the process have benefited from the discipline and sagacity that design can offer? Quite possibly, though only if it was accompanied by the necessary veterinary and zoological knowledge. As to whether someone who describes him or herself as a designer would need to be involved, that is a different matter.

2 What is a designer?

All men are designers. All that we do, almost all the time, is design, for design is basic to all human activity.
— Victor Papanek[1]

Every aspect of Edward Teach's appearance was calculated to make him look as fearsome as possible. Heavy coats, sturdy boots and imposing hats accentuated his height. His face was obscured by the straggling beard that inspired the nickname by which he was renowned as one of the most brutal and successful pirates of the early eighteenth century, 'Blackbeard'. Whenever his ship raided another vessel, Teach struck dastardly poses with several braces of pistols slung on his shoulders and lighted matches sizzling under the brim of his hat.[2] In his 1724 book *A General History of the Robberies and Murders of the Most Notorious Pyrates*, Charles Johnson described him as 'such a figure that imagination cannot form an idea of a fury from hell to look more frightful'.[3] The finishing touch was the macabre symbol of a human skull above a pair of diagonally crossed bones on his ship's flag.

Other pirates were adding those motifs to their flags at the time, for the same reason: to frighten their victims into surrendering. Pirates had deployed flags as part of their terror tactics for centuries, but had done so individually. When the Scottish buccaneer William 'Captain' Kidd was wreaking havoc on the high seas in the late 1600s, he often befuddled his prey by flying a French flag, hoping to lull them into a false sense of security until his ship was ready to attack. Henry Avery, an English pirate of the same era, preferred to dazzle his victims with a faux aristocratic flag bearing four gold chevrons. But by the turn of the eighteenth century, after years of war, Europe began a peaceful era, colonial trade flourished and New York emerged as a lucrative black market for contraband. Piracy became so profitable that the cannier practitioners, like Blackbeard and his Welsh counterpart, Bartholomew 'Black Bart' Roberts, were eager to operate as efficiently as possible. They treated their trade as a serious business and ran their ships accordingly, striving to complete raids swiftly, preferably without wasting valuable ammunition or incurring casualties. Terrifying the crews of the ships they attacked into speedy surrender was a prudent strategy, and flying a flag that declared how brutal and remorseless they could be was an inspired way of doing so.[4]

Maritime historians do not know exactly when the skull and crossbones was chosen as the collective symbol of piracy, or why the flag was called the 'Jolly Roger', but its imagery is self-explanatory. Having signified death in many cultures for centuries, that motif would have been instantly recognizable to sailors on either side of the law, regardless of where they came from. The first recorded sighting was by John Cranby, captain of the British Navy ship HMS *Poole*, in July 1700 when he spotted a skull and crossbones on the flag of a French pirate vessel captured off the Cape Verde islands in the Atlantic Ocean.[5] News of the bloodcurdlingly effective new flag would have spread swiftly in a cosmopolitan trade like piracy, and more sightings were soon made.

As well as communicating its intended message clearly, the Jolly Roger was flexible enough to be customized. The flag of the French ship pursued by HMS *Poole* sported an hourglass as well as a skull and crossbones to suggest that time was running out for its victims. Other pirates added daggers, skeletons or spears. One of Black Bart's flags bore two human skulls, each representing an enemy against whom he had sworn vengeance.[6]

Centuries later, the story of the Jolly Roger's origins reads like a textbook example of modern communication design, and the symbol of terror adopted by those lawless, mostly illiterate pirates seems like a precursor of today's corporate logos. It tells us exactly what its owners want us to think of them – in the pirates' case, that they were very scary – just as BP's sunflower symbol tries to convince us that it is a sensitive, environmentally aware company, despite the calamitous Deepwater Horizon oil spill, and the 'aristocratic' shield with a crown and cross printed on Prada's bags and boxes is a reminder that it has been making luxury goods for over a century.[7] However much more money, research, time and resources were invested in the development of those contemporary corporate motifs, they fulfil the same function as the skull and crossbones did in the early 1700s.

When most people think of 'a designer', they do not picture Blackbeard or Black Bart, but one of the larger-than-life design-heroes (rarely heroines) flashing carefully cultivated scowls on magazine covers. They might imagine someone like the redoubtable German graphic designer Jan Tschichold, who, as head of design at Penguin Books in London during the late 1940s, drafted the 'Penguin Composition Rules' and insisted that everyone in the company adhere to them. Suspecting Penguin's printers of prioritizing quantity over quality (they were paid by the keystroke), Tschichold prowled around the presses to check that they were observing his edicts. If any of them dared to challenge him, he would exaggerate his German accent and pretend not to understand what they were saying. When the novelist Dorothy L. Sayers objected to the decorative asterisks he had added (in contravention of his own rules) to her translation of Dante's *Inferno*, Tschichold retorted: 'The master is permitted to break the rules, even his own.'[8]

Equally draconian was the Danish furniture designer Verner Panton. When the Swiss industrialist Rolf Fehlbaum asked him to design the interior of his home in the late 1960s, Panton saw the project as a chance to put his colour theories into practice. 'Verner said: "We'll do a black room, a red room, a golden room and an orange room,"' recalled Fehlbaum, who soon discovered that the designer did not simply envisage the walls and floors being in those hues, but everything else too: every stick of furniture and whichever objects were likely to be used there. When Fehlbaum asked jokingly whether it would be permissible to move, say, a cup from one room to another, Panton was stupefied. Why would he want to ruin the colour scheme?[9]

Then there was the imposing personality of Finland's most famous twentieth-century architect and furniture designer, Alvar Aalto. Passengers on Finnair flights taking off from Helsinki in the 1950s and 1960s grew accustomed to leaving later than expected thanks to his tardy arrival. Reputedly the only person for whom Finnair routinely delayed its flights other than the country's president, Aalto invariably mounted the aircraft steps several minutes after the plane was due to depart. At first, the cabin crews assumed that he was not particularly punctual, but soon realized there was another explanation. Before a couple of his 'late' embarkations, Finnair employees spotted his car being driven around the airport before he boarded the aircraft. Having come to the airport in plenty of time, Aalto had instructed his driver to delay their arrival, so he could make a grand entrance.

Entertaining though such stories are (unless you were

one of Penguin's harassed printers, Panton's clients or Finnair's stewardesses), the image of the omnipotent design-hero has always been illusory.[10] Not only do designers, even the most famous ones, generally work as part of teams, which have to co-exist with other teams, their power to determine the outcome of their work is subject to numerous constraints. Any number of factors can impede the progress of a design project: the unwelcome intervention of colleagues, clients or suppliers; budgetary restrictions; human error; and sudden changes that the designer could not have anticipated, and cannot control. The completeness with which any individual designer's vision will eventually be realized is usually determined by a combination of their strength of character, persuasive powers and luck. All in all, their role is more ambiguous than is generally supposed, and is not restricted to professionals, as Qin Shihuangdi, Black Bart and countless other anonymous designers have proved. Anyone who conceives or implements the process of change we call 'design' is entitled to describe themselves as a designer, not just those who have been trained, or are paid for doing so.

Yet the myth of the dictatorial 'design-hero' has proved surprisingly enduring. It dates back to ancient Greece, when potters identified their urns by inscribing them with their own maker's marks. Typically, they sought credit for work that they considered to be aesthetically pleasing, not because it displayed the intellectual ingenuity or aptitude for change that is now expected of design. Most other design exercises were unrecognized at the time, went unrecorded and are now long forgotten. The few we are aware of are generally remembered for different reasons: as Blackbeard and Black Bart are for being notorious thieves who were lucky enough to be romanticized as plucky buccaneers by authors like Daniel Defoe, and a succession of movie stars from Douglas Fairbanks and Errol Flynn to Johnny Depp.[11]

The same applies to Nicholas Owen, a sixteenth-century carpenter in Elizabethan England, known as 'Little John', who was one of the first craftsmen to be noted for the conceptual strength of his work rather than for its beauty or finesse. As a devout Catholic in a Protestant regime, Owen lived in fear of religious persecution, and put his carpentry skills to good use by constructing 'priests' holes' inside the walls of houses to conceal fellow Catholics. Sometimes they had to stay there for days at a time, and Owen, who only ever worked at night and in deepest secrecy, added trapdoors, known as 'feeding traps', through which food could be delivered. In extreme circum-

A working model of the Manchester University Mark I computer in 1948

IBM System 360 mainframe computer in the 1960s

(left)

IBM 701 mainframe computer in the 1950s

Production of the IBM 701 mainframe computer in the 1950s

Apple iPad mini in 2013

stances, he would make a tiny hole in a wooden panel so that a hollow quill could be inserted and soup dripped into the fugitive's mouth. He was equally skilful at disguising the entrances behind pivoting floorboards or wall panels. Owen built dozens of 'holes' from the 1580s onwards, and may well have made more, but concealed them so cleverly that they have yet to be discovered. Eventually, he himself had to go into hiding and was arrested after being starved out of one of his lairs at Hindlip Hall in Worcestershire.[12] Believed to have given himself up to distract attention from an elderly priest who was concealed nearby, Owen died under torture in the Tower of London, having refused to betray his fellow Catholics.[13]

Other early 'designers' were accomplished tinkerers, whose design coups have, like Owen's, only been acknowledged because their principal achievements prompted historians to explore other aspects of their lives. Typical was Benjamin Franklin, the eighteenth-century US politician, political theorist and keen amateur inventor who devised the Franklin Stove to heat the air in a room to an even temperature. He also designed what are thought to have been the first bifocal lenses to help him to read in poor light, and a scissor-like contraption for retrieving books from high shelves.[14]

Similarly, Thomas Jefferson, the third president of the United States, added a revolving arm to the chair on which he wrote the 1776 Declaration of Independence. Jefferson was intensely involved with the design of his house, Monticello, near Charlottesville, Virginia, and installed several of his innovations there, including a dumb waiter hidden inside a fireplace. Full bottles of wine were hoisted up from the cellar on one side, and empty bottles sent down on the other. He also devised a clock to record the days of the week as well as hours and minutes.[15] Another supporting role in design history goes to the nineteenth-century British naturalist Charles Darwin, who built an early example of a chair mounted on wheels for the study of his home, Down House, in Kent.[16] Having replaced the legs of a wooden William IV-style armchair with a set of cast-iron bed legs mounted on castors, Darwin would roll around the room to inspect the rows of specimens laid out on the desks by his research assistants on a forerunner of the wheeled office chairs that are now used by millions of people every day.[17]

While these 'celebrity' tinkerers were finessing their inventions, the role of the designer was being formalized by industrialization. Once it became possible to make an object in huge quantities, it was necessary to ensure that each example was manufactured to identical specifications. The industrial design process was developed to fulfil that function, and the design profession invented to execute it. Charles Le Brun was one of the first designers in this sense, through his work at Gobelins,[18] followed by the sculptors who created delicate porcelain figurines for Meissen in early eighteenth-century Germany.[19] But the role was crystallized by Josiah Wedgwood in his Staffordshire potteries during the late 1700s.

Born into a local family of potters, Wedgwood had learnt his trade as a teenage apprentice before working with Thomas Whieldon, an older, more experienced ceramicist who had experimented with innovative ways of organizing the production process by allotting particular tasks to groups of workers.[20] Throughout his apprenticeship, Wedgwood proved to be unusually skilful at modelling, which involved shaping the pots, and he later supervised the training of the apprentice modellers in his own potteries, mostly local boys, such as William Wood, whose father had worked for Whieldon, and William Hackwood, who became his own most talented modeller.[21] Given that Wedgwood also chose the artists who decorated his ceramics, his personal taste was critical in determining the type of products made by his factories.

Fascinated by science, Wedgwood constantly experimented with new materials, glazes and production techniques, and discussed the results with fellow members of the Lunar Society of Birmingham, a group of intellectually vigorous local scientists and industrialists, including the chemist Joseph Priestley and the engineer James Watt.[22] He also took a lively interest in art and architecture, not least as possible sources of ideas to be applied to his products. When Wedgwood began in business, the most popular decorative style was ornate rococo, but he realized that it would soon be supplanted by the restrained neoclassical aesthetic of the fashionable Scottish architect Robert Adam. In 1763, Wedgwood introduced a richly glazed, cream-coloured dinner service made from unusually fine earthenware in simple, elegant shapes inspired by Adam's architecture. When Queen Charlotte, the wife of King George III, placed an order, he christened it 'Queen's Ware' in her honour.[23]

Five years later, Wedgwood accepted the most ambitious commission of his career: to make the Green Frog Service for Catherine the Great, Empress of Russia, as a fifty-person dinner service destined to be used in a summer palace built on a frog marsh near St Petersburg. Catherine stipulated that each of the pieces, all Queen's Ware, should depict a green frog and a scene from British life. Wedgwood dispatched artists and illustrators all over the country to sketch and paint over a thousand traditional vistas of forests, rivers, lakes, hills and castles, as well as glimpses of industrial Britain, such as canals and ironworks. Most of them worked on paper, but one artist was equipped with a camera obscura, a forerunner of the camera.[24]

As Wedgwood's reputation rose, he was able to persuade famous artists such as George Stubbs, Joseph Wright and John Flaxman to work for him.[25] They were described as 'designers', but their role was mostly limited to decoration, by drawing picturesque scenes or figures to be reproduced by the modellers. Together with Wedgwood, the modellers were responsible for critical design decisions such as the choice of shapes, materials, finishes and how the products were to be made.[26]

The vitality of enterprises like Josiah Wedgwood's seemed so thrilling to fashionable Londoners that they took tours of the bustling factories in northern England and the Midlands. By the early 1780s, painting his plain ceramics was regarded as an elegant accomplishment for society ladies alongside singing, playing the piano and needlepoint. Some 'artistic' socialites pleaded to be allowed to design pieces that would be manufactured by his factories. Wedgwood permitted a few of them to do so, including Diana Beauclerk, daughter of the Duke of Marlborough, and Emma Crewe, whose wealthy mother was among his best customers. Ever the astute marketeer, he may have been swayed by their publicity value as much as their talent, but the work of one socialite, Lady Elizabeth Templetown, a stylish amateur painter married to a royal courtier, Baron Templetown, proved to be surprisingly successful. She specialized in depicting sentimental, domestic scenes, inspired by ancient Greek mythology and Goethe's poetry, mostly of mothers nursing their children or contentedly engaged in domestic tasks. Her pieces for Wedgwood sold so well that rival manufacturers raced to copy them, giving the vivacious Lady Templetown a strong claim to be the first female industrial designer.[27]

The fashion for industrialization soon ebbed. By the early 1800s, millions of workers and their families had exchanged rural poverty for urban squalor by abandoning the countryside for better-paid, but often dangerous jobs in filthy, noisy factories. The socialites and intellectuals who had once signed up for factory tours regarded manufacturing as dirty, soulless and destructive, rather than exhilarating. On the rare occasions that industrialization appeared in nineteenth-century art and literature, it was generally demonized. The story of Victor Frankenstein, the idealistic scientist who is terrorized by his own invention, the 'Creature', in Mary Shelley's novel *Frankenstein*, is a morality tale warning against unfettered faith in science and technology.[28] When the novel was adapted for the London stage in the 1820s, Shelley's sensitive text was rehashed as an anti-industrial Gothic melodrama, and her 'Creature' reinvented as the 'Monster'.[29] A rumour that a railway was to be built near their sedate country town provoked a similar response from Dorothea Brooke and her neighbours in George Eliot's *Middlemarch*, set in the early 1830s.[30] Similarly, the heroine of Elizabeth Gaskell's novel *North and South*, Margaret Hale, was horrified when her family moved from a pretty village in southern England to the remorseless bustle of Milton, a brash industrial city in the tellingly named northern county of Darkshire.[31] Artists lost their early enthusiasm for industrial commissions, leaving manufacturers to employ draughtsmen to draw the specifications of their products, and engineers or modellers to interpret them for production. But the new designer-draughtsmen tended to be poorly paid, sparsely trained and exerted little influence over their employers. Mostly, they copied historic shapes and motifs from books, and the quality of their work was often questionable. The title page of the British architect A. W. N. Pugin's 1836 book *Contrasts* sported a spoof advertisement for 'Designing taught in six lessons: Gothic, Severe Greek and mixed styles' plus one for 'An Errand Boy for an Office who can design occasionally'.[32]

Politicians including Benjamin Franklin in the United States, Robert Peel in Britain and the French social reformer François Alexandre Frédéric, Duke de La Rochefoucauld-Liancourt, lobbied their governments to improve the training of designers and to promote their wares. France led the way by founding a national network of Écoles des Arts et Métiers to educate designers and engineers, and by showing off its manufacturers' output in a series of trade fairs, the Expositions Nationales des Produits de l'Industrie Agricole et Manufacturière, or 'Expos' for short.[33] Other countries followed suit by opening their own design schools and trying to outdo one another by staging ever bigger, more ambitious Expos, culminating in the first of the World's Fairs, the 1851 Great Exhibition in London.[34] Some six million people flocked to the Crystal Palace, a specially built glass structure so immense that it enclosed several fully grown trees, to marvel at more than a hundred thousand objects, including the world's largest diamond, a prototype of Samuel Colt's Navy revolver, a 'sportsman's knife' with eighty blades, a robotic 'Man of Steel' and the first public toilet.[35] The profits from ticket sales were used to buy land in nearby South Kensington to construct schools and museums, including the National Art Training School, later renamed the Royal College of Art, and the Victoria & Albert Museum, which opened in 1857 with many of the Crystal Palace's exhibits in its collection.[36]

The tide turned again when the artist William Morris, art critic John Ruskin and fellow members of the Arts and Crafts Movement dismissed industrialization as crassly commercial and championed a return to traditional craftsmanship. Such criticism stemmed partly from snobbery. The chief consumers of factory goods were the newly affluent middle classes, who were eager to acquire cast-iron and papier mâché replicas of the intricately crafted wares they associated with the aristocracy.[37] By then, the handcrafted work beloved of Morris and Ruskin tended to be either an indulgence of the rich or the last resort of the destitute, who had no choice but to make things themselves because they could not afford to buy them ready-made.

Morris practised what he preached at his decorating firm, originally named Morris, Marshall, Faulkner & Co and later Morris & Co, by collaborating with artist friends such as Edward Burne-Jones and Ford Madox Brown on the design of furniture and furnishings to be made by carefully selected artisans and workshops. 'The Firm', as they called it, then sold their wares.[38] Other designers chose to work within industry in the hope of raising standards there, including Christopher Dresser, who abandoned his original career as a botanist to do so. Rather than being employed by a particular company, like most of his peers, or setting up in business, as Morris had done, Dresser produced designs for furniture, wallpaper, ceramics, textiles and other products for different manufacturers from his studio. He also wrote extensively on design, and published an influential study of Japanese aesthetics. His knowledge and passion were reflected in his work, the metalware in particular.[39] Simply styled and aesthetically pleasing, his products reflected a thorough understanding of the economics of mass production and the skills of the workers who had made them.

As industrialization expanded, more designers were hired, the demands on them became more onerous and their roles more rigid. Design teams grew larger and more hierarchical. The historic division of responsibility between design and other disciplines, such as engineering, was formalized, and the friction between them escalated as the competing camps jostled for power. The quality of design education improved, and the influence of progressive schools, like the Bauhaus in Germany, infused designers with Constructivist fervour, encouraging them to be more intellectually ambitious about their work, and to think beyond the commercial demands of design, by treating it as a tool for effecting social and political change and a means of self-expression.

Some designers sought to achieve these goals within a commercial context, as Dresser had done. The design team headed by Peter Behrens during the early 1900s at AEG, the German domestic appliance maker, established a template for thoughtful, enlightened corporate design, which was subsequently applied by Jan Tschichold at Penguin, Eliot Noyes at IBM, Dieter Rams at Braun and more recently by Jonathan Ive at Apple.

Other designers cast themselves as iconoclasts, including László Moholy-Nagy, a charismatic Hungarian artist who was a committed Constructivist before serving in the army during the First World War. After the war, he fled Hungary to join Dadaist groups in Vienna and Berlin, where he produced a series of *Telephone Paintings* made by phoning instructions to a sign factory, which executed them. During the mid 1920s, Moholy became an influential teacher at the Bauhaus, where he sported red factory overalls to symbolize his faith in industry and the students nicknamed him 'Holy Mahogany'. He then returned to Berlin, to experiment with photography and film with his compatriot György Kepes. In the mid 1930s, he and Kepes left Nazi Germany to seek refuge in the Netherlands, Britain and eventually the United States.[40] Wherever they went, Moholy insisted on taking the Light Space Modulator, an electrical contraption he had assembled to create pools of light and shadow for him to study. It looked so strange that he described it variously as a 'robot', a 'fountain' and 'hairdressing equipment' to get it through customs.[41] To Moholy, the practice of design was 'not a profession but an attitude' and design itself was a holistic medium infusing every area of society. 'Ultimately all problems of design merge into one great problem: "design for life",' he wrote. 'In a healthy society this design for life will encourage every profession and vocation to play its part since the degree of relatedness in all their work gives to any civilization its quality.'[42]

After Moholy's death in 1946, Kepes continued their research and disseminated their ideas in his books on visual theory.[43] He also propagated them as a teacher: first at Brooklyn College, where his students included Saul Bass, who would introduce their avant-garde theories to a mass audience in the title sequences he devised for the films of Alfred Hitchcock, Stanley Kubrick and Martin Scorsese; and later at the Massachusetts Institute of Technology, where his colleagues included Muriel Cooper. There, Kepes founded the Center for Advanced Visual Studies, which became a role model for art and technology programmes all over the world. The experiments begun by him and Moholy in Berlin during the 1920s have a lasting legacy in the torrent of digital imagery that fills our lives today.[44]

Yet despite their best efforts, and those of other visionary designers like Cooper and Bucky Fuller, public perceptions of design were dominated by the flamboyant personas of the two most visible designers in both halves of the twentieth century: both freelance guns-for-hire like Christopher Dresser, both French and both delighted to play the part of traditional design-heroes.

First came Raymond Loewy, who left France on a ship bound for the United States in 1919 with $50 in savings and a Croix de Guerre medal for his military service during the First World War. After starting out as a window designer for Macy's and Saks and as a fashion illustrator for *Vogue*, he reinvented himself as an industrial designer. Loewy was responsible (or he and his team were) for the design of the Greyhound bus, the Lucky Strike cigarette packet, a Coldspot refrigerator, the Coca-Cola bottle and logos for Shell and Exxon, among other things. In 1949, he became the first industrial designer to appear on the cover of *Time* magazine. A line drawing of Loewy's face was encircled by some of the hundreds of products he had designed.[45]

An imposing figure, dapperly suited and deeply tanned with a neatly trimmed moustache, Loewy was a relentless self-publicist who conducted countless interviews and published several books on his life and work, bragging about his achievements and shamelessly dropping the names of famous friends and clients. Typical was an anecdote about President John F. Kennedy telling his secretary that he and Loewy 'were not to be disturbed' in the Oval Office after inviting him to the White House to discuss the redesign of Air Force One.[46] Loewy also relished the tale of his housewarming party at the spectacular modernist home in Palm Springs he had commissioned from the Swiss-born architect Albert Frey. No sooner had his neighbour, the movie star William Holden, fallen into the pool fully clothed, than the singer Tony Martin followed suit, and then their host.[47] 'Asides from wanting to make a living, we were all nice fellows,' he said of himself and his fellow industrial designers. 'Outré at heart and simple enough to believe that by improving a product functionally, safely, qualitatively and visually, we were contributing something valuable to the consumer, his sense of aesthetics and to the country.'[48]

If Loewy created the role of the super-prolific, mediagenic designer-for-hire, Philippe Starck perfected it. He made his name in the 1980s as a postmodernist prankster, designing a chair for Café Costes in Paris as a playful pastiche of an early 1900s Viennese coffee-house chair with three legs rather than the usual four.[49] At the time, Starck said that the waiters would trip up half as often on a chair with one back leg, not two.[50] Subsequently, he made three legs a signature of his chairs, despite the complaints of the luckless souls who tumbled off them on to the floor. Starck continued in a similar vein by producing plastic versions of Louis XV-style chairs, stools shaped like garden gnomes and lamps mounted on spookily realistic replicas of Beretta pistols and AK-47 assault rifles. A burly, *mal barbu* figure who bore a distinct resemblance to Desperate Dan, the pie-scoffing hero of a cartoon strip in the British children's comic the *Dandy*, he courted the media by speaking in franglais sound-bites. Once, Starck boasted of having designed a chair in the few minutes between an aircraft seat-belt sign going on and off. He also claimed to have a Harley Davidson waiting for him in most of the cities where he worked. For years, Starck claimed to be bored by design, and at times he designed as if he was, which was a shame, as his best work was spirited and witty. He then attempted to reinvent himself as a champion of sustainable design, despite also accepting the role of creative director of a space tourism venture and producing yet more variations of his best-selling (but, sadly, non-biodegradable) plastic Louis Ghost chair.[51]

He and Loewy were more design-showmen than design-heroes, but they were cast in the indomitable mould of Panton and Tschichold, and did little to disrupt conventional expectations of designers, especially in their relish for the commercial aspects of their work.[52] 'I once said that industrial design keeps the customer happy, his client in the black and the designer busy,' Loewy noted in his monograph-cum-memoir *Industrial Design*. 'I still feel this is a good maxim.'[53]

If he and Starck typified old-school designers, how do today's designers differ from that stereotype? Many of them do not, because they have chosen similar ways of working, even if they are less wealthy and less famous than those Gallic show-offs. A critical difference is that they now have many other roles to choose from. Designers can pursue environmental, political and humanitarian causes by casting themselves as activists and adventurers like Nathaniel Corum, or social reformers like Hilary Cottam.[54] They can act as auteurs by deploying design as a means of self-expression in conceptual projects, whose function is neither practical nor commercial, but one of intellectual enquiry, or join the growing group of critical designers, who use their work to critique design culture, as Roland Barthes and Jean Baudrillard did in their writing. Or they can immerse themselves in specialist fields such as medical research, supercomputing and nanotechnology, which were once the preserve of scientists, not designers.

Whether their efforts are driven by empathy, outrage or curiosity, digital technology is likely to be an indispensable part of their practice. Take Emily Pilloton, who left a job in commercial design in 2008 to form Project H, a volunteer network of humanitarian designers, from the dining table of her parents' home in California with $1,000 of savings. By the end of year one, Project H (the 'H' stands for humanity, habitats, health and happiness) had established local 'chapters' of volunteers in Johannesburg, London, Mexico City and six cities in the United States to provide learning tools for schools all over the world and clean water to rural communities in Africa. Pilloton started Project H on her own, armed with a laptop, on which she rounded up volunteers, funders and collaborators. She then raised awareness of its work by posting about it on social media networks and blogging on empathetic websites, before repeating the exercise for Studio H, an experimental high-school design course that started out in one of the poorest rural areas of North Carolina.[55]

For a social design group like Cottam's Participle, which deploys design strategically, digital technology is essential both as a communications tool and as a means of analysing huge quantities of complex data with the precision required to deliver smarter public services.[56] Rather than risk wasting resources by providing standard packages to everyone, as traditional social services dealing with, say, elderly people or the long-term unemployed have done, Participle marshals its database to identify what type of support each individual needs, then uses it to monitor their progress and, if necessary, to make modifications. Without such sophisticated analysis, it would have to employ so many people to crunch the same information that its new services would be impracticable in terms of time as well as cost.[57]

Like Project H, Participle sets its own agenda. Rather than waiting to be commissioned by political bodies or charitable foundations, it identifies the areas where it wants to work, and drafts a plan of action based on focused research. It then pitches the plan to prospective funders and partners, before choosing the ones that promise to be the most productive collaborators. The same entrepreneurial zest is evident in other expanding areas of design[58] and has proved decisive in enabling designers to seize the autonomy that had long eluded them when they were restricted to commercial roles and subject to instruction from senior colleagues or clients.[59]

Equally important is a close rapport with other disciplines. Participle's projects are usually led by a designer and adhere to the conventional structure of the design process, but the participants are as likely to be economists, statisticians, ethnographers, anthropologists, psychologists, computer programmers, business specialists or social scientists (like Cottam herself) as they are to be designers. IDEO adopts a similar approach when applying design thinking in the commercial sphere by assembling teams of designers, behavioural scientists, engineers, psychologists and other specialists.[60] Designers also need to draw on the knowledge and experience of people with different skill sets for specific projects, for example when responding to new directions in technology. Some of the most interesting innovations in digital books have been executed, not by graphic designers, but by animators and film-makers.[61] The same applies to designers' efforts to help us to live more responsibly by developing a sustainable society. Much of this work would be impossible without specialist scientific input, which is why the American landscape architect William McDonough teamed up with the German chemist Michael Braungart to develop the 'Cradle to Cradle' system of sustainable design and production.[62] If designers are to continue to grapple with such challenges, they will be likelier to do so as team players, than as go-it-alone 'design-heroes'.

More 'design-heroines' should be joining them in future.

Historically, design has been a boy's club, and a white boy's club at that, which is why design history books tend to be filled with white male faces, plus a few Japanese ones. Even the supposedly progressive Bauhaus limited women to studying ceramics or weaving during its early years. Successful female designers were relatively rare until recently, and many of them belonged to couples, like the American product designer Ray Eames and her husband, Charles, and the German interior designer Lilly Reich and her lover, the architect Mies van der Rohe.[63] Unfortunately, they were often overshadowed by their male partners. Poor Ray Eames suffered the indignity of being introduced to the audience of the NBC television show *Home* in 1956 with a condescending: 'This is Mrs Eames and she is going to tell us how she helps Charles design these chairs.'[64]

When the twenty-three-year-old Charlotte Perriand arrived at Le Corbusier's architectural studio in Paris in 1927 hoping to persuade him to offer her a job, he rebuffed her with a brusque: 'We don't embroider cushions.' Days later, he saw a room set designed by Perriand at the Salon d'Automne exhibition in Paris, and hired her on the spot.[65] She worked in his studio for ten years, becoming the lover of his cousin and collaborator Edouard Jeanneret, before embarking upon a successful independent career. Yet Perriand was a rare exception, as was Charles Harrison, who became the first African American designer to win a Lifetime Achievement Award from the Smithsonian's Cooper-Hewitt National Design Museum in New York in 2006. When he had applied for a design job with the retail group Sears Roebuck fifty years before, Harrison was told that the company had an unwritten policy against hiring black people. He was taken on by a commercial design consultancy, where he executed several assignments for Sears. Five years later, Sears offered him a job, and Harrison worked there for over thirty years, becoming chief designer and developing some of its best-selling products.[66] Like Perriand, he set an encouraging precedent, and the design profession has since become more diverse in terms of gender, ethnicity and geography. Design's expansion into new terrain should accelerate future progress, not least because 'outsiders' tend to flourish in new fields where there are fewer barriers to entry and no established order to close ranks against them.[67]

As the profession opens up, so will the design process. One catalyst is the growing popularity of open source development, which was pioneered by Dennis Ritchie and his peers in the 1970s to enable people to scrutinize each stage of a design project as it evolves, and to critique it.[68] Another is that we 'civilians', the people who use designers' work, are becoming less inclined to allow them to play 'the master' as Jan Tschichold did in his day, and are increasingly eager to design for ourselves, either by making things from scratch or by customizing them. Maker Faires have sprung up all over the world, where DIY designers show off latter-day versions of Benjamin Franklin's stove and Charles Darwin's wheeled chair.[69] No sooner did the first iPhone go on sale in 2007 than hackers found ways of devising unauthorized apps to download on to it. Apple eventually bowed to public pressure by allowing them to be sold on its Apps Store, keeping a hefty percentage of the proceeds as a quid pro quo. Soon, it was selling over a billion apps every month, mostly the work of self-taught programmers.[70]

Nor is the zest for customization likely to ebb. On the contrary, it will accelerate with the progress of digital production technologies, like three-dimensional printing, which are so fast and precise that products can be made individually or adapted to meet each person's needs at no extra cost. Eventually, every neighbourhood and village could have its own 3D printer, which will operate like an old-fashioned blacksmith's forge by making new things for local residents and businesses, and repairing old ones.[71] Ensuring that these technologies are put to good use is an important challenge for designers in their traditional role as agents of change. To discharge it successfully they need to redefine their relationship with the rest of us by enabling us to participate in the design process, and to do so constructively.

If more and more people cast themselves as designers, where does this leave the professionals? Will they disappear? Or will their influence be gradually eroded? Not if they prove their worth. In this respect, design is not unlike psychology. Lots of us like to think of ourselves as self-taught psychologists, and often use rudimentary psychological techniques when coming to instinctive conclusions about other people's actions or motives. If we are lucky and observant, our judgements may be correct, but surely they would be more perceptive if we had studied psychology, or were able to draw on the knowledge, discipline and experience acquired from years of professional practice. Much the same can be said of design, just as long as the professionals can match the originality and resourcefulness of Blackbeard, Qin Shihuangdi, Nicholas Owen and other 'accidental' designers.

3 What is good design?

Tasteful rubbish is still rubbish.
— Reyner Banham[1]

For one of the most popular lecturers of his time, William Morris was a surprisingly nervous public speaker. During the late 1800s, he aired his opinions on art, design, politics, education, the nature of civilization, the history of Byzantine textiles and the palaces of the Assyrian kings in hundreds of lectures throughout Britain. Morris prepared meticulously for each one, writing out the words in lined exercise books and revising them doggedly, yet his family and friends watched anxiously as he stood at the lectern, toying with his pocket watch and shuffling from foot to foot. Afterwards he shared his concerns in letters to his wife Janey, worrying that the people in the audience had not understood him, or were somehow dissatisfied.[2]

Morris was more confident when it came to the content of his lectures. Take 'The Beauty of Life', a forbidding subject that he addressed in 1880 with a whistle-stop tour of art history, a concise comparison of Gothic architecture and literature, followed by scathing analyses of the cultural consequences of the Industrial Revolution and 'the (so-called) restoration of St Marks in Venice'. He also distilled his theme into what he described as 'a golden rule that will fit everybody', namely: 'Have nothing in your houses that you do not know to be useful and believe to be beautiful.'[3]

'Useful' and 'beautiful'. Those words were Morris's contribution to a debate that has preoccupied designers, design theorists and design historians ever since. These days, the title of his lecture would be something like 'Good Design' rather than 'The Beauty of Life', though defining exactly what that is would be even more difficult now than it was in Morris's era.

The design curators at the Museum of Modern Art in New York devoted a decade of exhibitions to what they considered to be 'the best modern design . . . available to the American public' starting in 1938. Those shows inspired the foundation of the

Good Design Awards at the Chicago Athenaeum Museum in 1950.[4] The same theme was the leitmotif of the work of the various government-funded bodies that championed design in post-war Europe: 'good design' in Britain, *bel design* in Italy and *gute Form* in Germany. If you Google 'good design' today, you will discover the latter-day version of the Chicago awards as well as dozens of other design prizes, magazines and websites. In every incarnation, 'good design' has been a combination of different qualities, but what those qualities are, their relative importance and how they respond to each other has changed significantly over time. Where do they all stand now?

One essential tenet of good design, which has existed throughout the ages, is that no design exercise can be deemed worthwhile unless it fulfils its function and does so efficiently. Plato said as much in 390 BC by stating that the 'virtue and beauty and rightness of every manufactured article, living creature or action is assessed only in relation to the purpose for which it was made'.[5] In other words: is it fit for purpose or, as Morris put it, useful? Not that it would suffice for it to be useful in theory, because good design must also be useful in practice, or simple to use.

It is frighteningly easy to think of examples of designs which are not: over-complicated phones; malfunctioning ticket machines; uncomfortable chairs; illegible typefaces; supposedly recyclable materials riddled with toxins; unreliable cars that are prone to breaking down, usually at the least convenient moments; television remote controllers with nearly as many buttons as 747 jumbo-jet flight decks. We all risk coming across them every day, more than once if we are unlucky.

A chilling example of design which was not fit for purpose is the original version of the M-16, the US Army's standard service rifle during the Vietnam War. The North Vietnamese and Vietcong forces were equipped with the AK-47 assault rifle, which had been used by the Soviet Army since 1949 and was described by C. J. Chivers, a former US marine officer and war correspondent, in his book *The Gun* as 'the most abundant and widely used rifle ever made'.[6] The story of the AK-47's design was romanticized by the Soviet authorities for propaganda purposes, but the end result was named after Senior Sergeant Mikhail Kalashnikov, who was wounded in 1941 while serving as a tank soldier and, according to the official version of events, developed a prototype in a local workshop during his convalescence. After years of testing and numerous modifications, the AK-47 became a model modern gun. Lighter and more reliable than existing assault rifles, it was capable of firing at a faster rate at close quarters using such small bullets that each soldier could carry more of them. The AK-47 was made from noisy moving parts that rattled alarmingly but were tough enough to cope with extremes of heat and cold, as well as rough treatment in turbulent conditions. Not only was it an asset to the North Vietnamese and Vietcong troops during the Vietnam War, but some of the earliest models, manufactured in the 1950s, have proved so resilient that they are used to deadly effect in war zones today.[7]

As the conflict in Vietnam escalated during the early 1960s, the US Army sent more troops there and stockpiled weapons. In 1963, it placed an order for over one hundred thousand M-16s, hoping that the new gun would be superior to the AK-47, but it was not. Chivers argues that the testing of the M-16 was rushed, and key decisions about its final specifications were taken by people with little or no field experience. The resulting rifle was accurate, often more so than the AK-47, but flawed in other respects. Critically, it was prone to jamming after being fired, and when that happened, removing the empty shell was cumbersome and time-consuming. This posed grave problems for US servicemen in Vietnam as they fought the North Vietnamese and Vietcong armies, both of which were armed with more efficient AK-47s. Chivers cites one marine who wrote to the local newspaper in his hometown describing a battle near Khe Sanh in the spring of 1967: '"Believe it or not, you know what killed most of us? Our own rifle."' A fellow marine, Gunnery Sergeant Claude Elrod, found an AK-47 beside the corpse of a North Vietnamese soldier and used it for most of the rest of the time he served in Vietnam. When his colonel demanded to know why he was carrying a Soviet weapon, Elrod replied: 'Because it works.'[8]

That is why the AK-47 passes the 'fit for purpose' test, and the early version of the M-16 did not. If a rifle is not reliable enough to be depended upon in battle, how can it be well designed? Similarly, how can a signage system be considered good design if it is not legible? A phone, if it is too complicated to be used easily because its operating system is so befuddling? Or a chair, if it is uncomfortable to sit on for long? It does not matter if they have other merits: if those signs are rendered in beautiful typography; if the phone boasts more functions than anything else on the market; or if the shape of the chair looks spectacular. However impressive they may be in other respects, they do not do their jobs properly. As the artist Donald Judd wrote: 'If a chair . . . is not functional, if it appears to be only art, it is ridiculous.'[9]

Conversely it is possible for something that fails on other criteria to qualify as good design, providing it is useful. Take Google's logo. Design purists tend to loathe it, and it is easy to see why. The company's standard logo, which spells the word 'Google' in brightly coloured, wonkily shaped letters, looks infantile. The original was designed by Sergey Brin, one of Google's co-founders, when he and a fellow Stanford University computer science graduate, Larry Page, began the business in 1998. The following year they asked Ruth Kedar, a graphic designer and friend from Stanford, to refine it. She slimmed down the typeface and lost a jaunty exclamation mark from the end, but kept the gaudy colours and its playfully amateurish style. By then, Brin and Page had also introduced what Google calls 'doodles' by temporarily replacing the usual logo with customized versions to mark special occasions.

The first doodle appeared in 1998 when Brin and Page took time off to attend the Burning Man Festival in the Black Rock Desert in Nevada. Instead of posting an 'Out of Office' message on the home page of their website, they hinted at where they were going by drawing a little stick figure behind the second 'o' of Google, in a nod to the wooden effigy which is ceremonially burnt at the festival.[10] Google has since posted hundreds of doodles to celebrate everything from Thanksgiving, Halloween, Earth Day and St Patrick's Day, to lunar eclipses, the launch of the Large Hadron Collider, the Mars Rover landing, major sports events and the birthdays of Jane Austen, Andy Warhol, Charles Darwin and the ice-cream sundae. Each one is executed in the same twee style as the everyday logo. Jackson Pollock's name was written in the 'drips' of his paintings, and Isaac Newton's accompanied by an animated apple falling off a tree.[11]

Looking cringy is one of the doodles' functions. They are intended principally to make us like Google. Trying to do so by convincing us that Google is still the sort of cool, friendly company whose co-founders might slope off to the Burning Man Festival is a good start, not least because it is fiendishly difficult for a business to convey that impression, especially one as big and powerful as Google. The doodles achieve this by giving us something new to look at for the day, while suggesting the sort of things that Google is into. So what are they? Judging by

R. Buckminster Fuller with models of the geodesic dome at Black Mountain College in 1948

László Moholy-Nagy
in Chicago in 1945

György Kepes at the Massachusetts Institute of Technology in 1971

Muriel Cooper at the Massachusetts
Institute of Technology in 1977

the doodles, Google likes science, literature and contemporary art, without being so snooty that it doesn't also enjoy the World Cup and ice-cream sundaes. Critically, the doodles communicate all this by suggestion: always more persuasive than being explicit, because we are likelier to believe in something if we think that we discovered it for ourselves.

Oddly, the clumsy aesthetic of Google's logos, both the everyday one and the doodles, makes this seem plausible. It is only natural to feel sceptical about whatever Google or any other powerful multinational company tells us about itself. No one likes to be taken in by a corporate titan and its army of expensively paid advisors. That is why corporate identities seldom work, because we are instinctively suspicious of them, often sensibly so. Besides, we have become so expert at decoding visual symbols like logos that the more sophisticated they seem, the deeper our suspicions are likely to be. This is why the cringiness of Google's logos is so clever. How can something so gauche be manipulative? Google's corporate identity excels thanks to its aesthetic flaws, because they help rather than hinder its efforts to fulfil its function.

There are also instances when usefulness alone is not enough to produce good design, because we expect it to deliver something else too. Traditional book covers come into this category. Of course they must protect the pages of the book, but if that is all they do, they will not qualify as good design. A really well-designed jacket should also sum up what the book promises to deliver so compellingly that we cannot resist reading it. The same goes for film titles. Telling us the names of the cast and crew of the film simply covers the basics. Well-designed title sequences, like the ones devised by György Kepes's student Saul Bass in the late twentieth century, not only seduce us into wanting to see the film, but heighten the experience of doing so.[12] Before Bass, most film titles simply listed the cast and crew, and were projected on to closed curtains, which were not drawn until the action began. The animated sequence he devised for the opening of the 1955 drama *The Man with the Golden Arm* was so striking that its director Otto Preminger attached a note to the film cans insisting that the curtains were drawn before the projectionists screened the movie.[13] Bass designed dozens of other title sequences over the years. He began Hitchcock's 1958 thriller *Vertigo* with the camera zooming into a woman's face and then her eye, which spirals terrifyingly as blood soaks across the screen;[14] while the opening credit sequence of Scorsese's 1995 film *Casino* shows Robert de Niro's body falling helplessly through the macabre neons of the Las Vegas Strip like a Dantesque descent into hell.[15] By preparing us for what we will see on screen, Bass's titles make us more emotionally responsive to the film, and more perceptive about it.

The 'something else' that elevates his cleverly conceived and executed film titles into not merely good, but great, design is their aesthetic impact, another important element of design throughout history. It is still immensely powerful, particularly as a medium of communication. Just as Bass's opening sequences for Hitchcock's thrillers convey foreboding and suspense, the Red Cross's symbol commands our attention and alerts us to danger, while insisting on respect and cooperation.[16] But one traditional element of design aesthetics is gradually becoming less important – beauty.

For a long time, beauty was deemed indispensable to good design, just like functionality. William Morris affirmed this in his 1880 lecture, and over half a century later, 'excellent appearance' and 'progressive performance' were chosen as the key criteria for inclusion in the first of the Museum of Modern Art's 'Good Design' exhibitions.[17] But Google's identity is not the only example of a successful design project whose looks are far from appealing. Another is the Post-it Note, which scores highly for usefulness and ingenuity, but poorly visually, at least in its original shade of urine yellow.[18] Nor is it pleasurable in any other way, in terms of smell, for instance, or touch.

Not that there is anything wrong with beauty in design: on the contrary, it can make our lives pleasanter, inspiring even. There is an argument that this is a field where design has taken over from art, partly because the visual elements of design have become so much more refined. Thanks to digital technology, designers in many disciplines can now work with greater expressiveness and precision. Imagine how miraculous an Apple iPad would have looked in the 1960s when a 'computer' meant an enormous room packed with millions of dollars' worth of noisy, cumbersome machinery; or how extraordinary the intricate digital imagery we see on their screens would have appeared.

This argument also reflects the changing role of art. For centuries, people sought beauty from art, but since the birth of modernism it has been expected to be challenging and provocative, by exploring subversive, frightening or ambiguous aspects of life that we struggle to understand, often because they defy rational analysis. If a painting or sculpture is simply beautiful, it can be difficult – though not impossible, as Cy Twombly and Lucio Fontana's gloriously sensual work proves – for us not to suspect it of being trite; whereas we are less likely to feel embarrassed about enjoying the graceful proportions and smooth surfaces of one of Fiskars' gardening tools, because we know that visually and sensually appealing though that trowel or rake is, it will also be useful. If it was not, we might suspect its beauty of being a deceptive trick, making us disinclined to feel quite so uninhibited about enjoying it.

Yet it is still possible for something that qualifies as good design on other criteria to fail because of its aesthetic shortcomings. An example is the London 2012 Olympic Games logo.[19] Functionally, it has several strengths: being recognizable, memorable, versatile enough to work in any medium – static in print, and animated on screen – and capable of being customized for use in different contexts, by changing colour or gaining new symbols. So far so good, except that, if you have actually seen the logo, you may think, as I do, that it is recognizable and memorable for the wrong reasons, because it looks so dreadful, with its ugly typography and clumsy shape. Some critics have likened it to a Nazi swastika, others to Lisa Simpson performing an obscene act.[20] And admirable though versatility is, what is the point if the logo looks equally dire in every incarnation? (An honourable exception being an unofficial version that appeared a few years ago on East London fly posters in which the numbers 2, 0, 1 and 2 were replaced by the letters S, H, I and T.[21])

Another tension inherent in design aesthetics is that such judgements are always subjective. After all, there may be some people who like the London 2012 logo and possibly the uriniferous colour of the Post-it Note. Deciding whether or not we find something visually pleasing is fairly straightforward. We tend to know instinctively, just as we sense intuitively whether we like or dislike the taste of food. Identifying why we feel that way is trickier. It is much easier to explain whether design succeeds functionally, because its performance is often, though not always, quantifiable. Analysing why we feel the way we do about how it looks is more complicated.

Yet there are some clear-cut cases. One is illustrated by the aesthetic differences between two typefaces that are included in the Font menus of most computers and are used in this book: Arial and Helvetica. Both are what are called sans serif fonts, which are ones without decorative flicks at the ends of the letters. At first glance they look very similar, so much so that each is often mistaken for the other. When Arial was introduced in 1982, it was generally seen as a copy of Helvetica, which dates back to 1957 and has since become a familiar sight on New York subway signs and corporate logos such as those of American Apparel, BMW, 3M and American Airlines.[22] But if you examine the characters in each font closely, the differences between them soon become apparent.

The American graphic designer Mark Simonson produced an excellent analysis of the two, which shows how much more refined Helvetica's detailing is than Arial's. The tail of the 'a' is gently curved in Helvetica, as is the first connection of the bowl to the stem, but not in Arial. Similarly, the top of the 't' and the ends of the strokes in the 'C' and 'S' are perfectly horizontal in the former, but slightly angled in the latter. He also noted that the stem of Helvetica's 'G' has a small spur at the bottom and a subtle curve flowing into it, whereas the stem of Arial's 'G' has neither.[23] In other words, the characters in Helvetica are more complex in structure than those in Arial. The distinguishing details are so tiny that you can only see them if you scrutinize magnified versions of each character as Simonson did. Only a handful of the millions of people who use either typeface will ever look closely enough to notice them. Yet it is these subtleties that make Helvetica a finer example of design than Arial. Functionally the two fonts are roughly equal, as both are admirably clear and easy to read, but aesthetically Helvetica is superior.[24]

The problem is that most judgements on the visual aspects of design are more muddled, and the impact of digital technology is making them more so. Consuming our daily diet of information and entertainment from the pixelated images on a computer or phone screen changes how we see the world, both on and off screen. Imagine looking at a stretch of street in real life, then seeing it on a photograph, cinema screen, television set, computer and phone. Think of how different it would appear in each one.

There are subtler influences too. Many of the objects we see and use in real life were designed digitally, with the designers sending computer files to printers or manufacturers. This process has a dramatic effect on how the finished pieces look. None of the 'blobby' shapes, the smooth ovals that appeared in product design during the 1990s, would have been possible without the precision of design software, nor would the futuristic forms of data visualizations, of the buildings of architects like

Rem Koolhaas, SANAA and Farshid Moussavi, or of the new genre of objects produced by 3D printing and other emerging technologies. Digital technology is not only changing what we see, but whether we like or dislike the result, often without our noticing.

One consumer products company set out to analyse this phenomenon by studying how people of varying ages and levels of computer-savviness responded to various objects. Half of them were designed digitally and the rest developed in the traditional way, but the participants were not told this. They were simply asked if they liked the look of each one. How the products were designed had no apparent bearing on the responses from people who rarely used computers. But it had a significant effect on the techier participants who spent lots of time on the Internet or playing video games. Almost all of the things they said they liked had been designed digitally. They were attracted to them instinctively, without realizing why.

The elusive influence of digital technology is one reason for our ambiguity about visually appealing design, but there are others too. One is fear of frivolity. In an era when the moral dimensions of issues like environmentalism and ethical concerns are increasingly important, it is easy to feel uncomfortable about enjoying the aesthetic side of design, because it can feel superficial to care about how things look. Equally problematic is our growing suspicion of beauty, especially the conventional variety, at a time when cut-price cosmetic surgery, jaw-dropping digital effects in movies and the digital retouching of photographs makes it increasingly difficult to distinguish artifice from reality.

Some designers have responded by challenging conventional stereotypes of beauty, often by giving their work a *jolie laide* quality, akin to the Japanese concept of *wabi-sabi*, which finds pleasure in imperfection or impermanence.[25] The work of the German product designer Konstantin Grcic often looks clumsy, even ugly, at first sight, but the longer you look at it, the more beguiling it becomes, as you sense the underlying logic of its blunt edges and awkward shapes. That is because Grcic does not begin the design process with a sense of what the end result will look like, but by visualizing how it should be used. He develops each object by building rough cardboard models and adjusting them, to make, say, a stool more comfortable to sit on or an espresso machine easier to operate. Once the model is completed, its dimensions are fed into a computer to refine the details.[26] The beauty of his products stems not from their looks, but from the warmth you feel towards them, once you have realized that the odd contours of the stool are perfectly positioned to enable you to perch on or to lean against it.

The Dutch designer Hella Jongerius achieves a similar effect by introducing intentional flaws to her objects – mis-matched colours and odd buttons on the upholstery of a sofa, and tiny irregularities on what we would expect to be the smooth surface of a dinner plate – to trick us into associating them with antiques, heirlooms and keepsakes that we have grown to love over time. She believes that this forges a stronger emotional rapport with the user than a 'perfect' object would do. Jongerius also puts considerable thought and effort into refining the tactile qualities of her work.[27] Powerful though these sensual aspects of design can be, articulating why we are drawn to them can be even more difficult than explaining why we are attracted to something visually, not least because the vocabulary for doing so is limited.

Other designers celebrate abstract qualities, which are related to the spirit of a design project and its impact on its surroundings. The British designer Jasper Morrison and his Japanese counterpart Naoto Fukasawa coined the term 'super normal' to describe this approach. A 'super normal' object is useful, appropriate, modest, robust, enduring and, as Morrison put it, should 'radiate something good'. Among the examples that he and Fukasawa chose were a Bic biro, a pair of Fiskars scissors and a Cricket disposable lighter.[28] They also included products developed by the Japanese industrial designer Sori Yanagi, whose work embodied many of the principles of the Mingei movement founded in Japan in the 1920s by his father, the writer Soetsu Yanagi. It celebrated the simple, solid virtues of the handmade everyday objects that had been overlooked in Japan since the Industrial Revolution. Such qualities have a similar effect to something that is aesthetically or sensually appealing, by giving us instinctive pleasure whenever we see or use them.

Another optional element of good design is originality, as many of Morrison and Fukasawa's 'super normal' products demonstrate. They are not new at all, yet their other design merits are unassailable. Nor is originality of value in itself, as Spencer Silver, a scientist working at 3M, discovered in the 1960s when he invented an unusual new type of glue. It was strong enough to stick something light, like a sheet of paper, to a smooth surface, but too weak to do so permanently, enabling you to peel it off whenever you wished. The problem was that 3M could not work out what to do with the glue, despite urging its research team to find a suitable – and marketable – function for it. In 1968 another 3M scientist, Art Fry, stumbled across a solution while singing in his church choir. He had placed slips of paper on the relevant pages of his hymn book, but was irritated to discover that they kept slipping out. Then he remembered Silver's formula for the sticky but not too sticky glue, and suggested that 3M should use it to produce a removable bookmark.[29] Before Fry's brainwave, Silver's glue had been pointlessly new. Putting it to good use as the sticky component of what was to become the Post-it Note transformed it into a valuable innovation and an inspiring example of good design – despite 3M's unfortunate decision to use that uriniferous yellow paper.

Optional though originality is to good design, it can be one of its most compelling elements. Design is often at its most seductive and most convincing when introducing us to the new. The first mass-manufactured chair. The first bicycle. The first motor car. The first aeroplane. The first high speed train. The first mainframe computer. The first personal computer. The first electric car. The first deep-sea drone. Many of the most thrilling episodes of design history have been filled with 'firsts'.

Not that those 'firsts' need to be conspicuous. A quietly impressive example is the series of screen-friendly typefaces developed by the British-born typography designer Matthew Carter for Microsoft during the mid 1990s. Up until then most fonts used on computers were originally designed to be read in print, and were not necessarily legible when seen on a screen, where each digital character is constructed from individual pixels, which are considerably more cumbersome than a fluid stroke of ink. Carter began by identifying the problems of replicating typographic characters digitally, then worked out how to resolve them. The characters that tended to be confused most often on screen were the letters i, j, l and the number 1. He made each one as simple in style as possible, with no superfluous details, and paid particular attention to the spacing between them when designing the first of the new fonts, the sans serif Verdana. The same principles were applied to its serif companion Georgia, which was even more problematic because of the bulk of the strokes at the ends of the characters, the numbers especially. Carter's solution was to exaggerate the difference between each number by varying their height. The 3, 4, 5, 7 and 9 in his serif font Georgia drop below the line, while the 6 and 8 rise above it.[30] Not only did these details make the new screen-friendly fonts clearer and more legible, they enhanced their aesthetic impact.

So far, good design equals something that must be useful, and may or may not be aesthetically pleasing or original. But there is another indispensable element alongside usefulness, which is best summed up as integrity. If the word strikes an old-fashioned moral tone, that is because it is intended to. Qualities such as honesty, clarity, sincerity, decency, soundness, incorruptibility and other components of integrity have been central to the debate on good design since Plato's *Early Socratic Dialogues*. Unless it has integrity, no design project can be deemed to be good, however useful, beautiful or innovative it may be. 'Good design enables honest and effective engagement with the world,' wrote the American philosopher Robert Grudin in his book *Design and Truth*. 'If good design tells the truth, poor design tells a lie, a lie usually related, in one way or another, to the getting or abusing of power.'[31]

Integrity extends to every aspect of design, starting with its purpose, and the objectives of whoever designed or made it. Chillingly efficient though the AK-47 is when it comes to fulfilling its function,[32] how can something which was designed to destroy human life be considered good design? It cannot.

Good design must also have symbolic integrity. Google's doodles do, despite their tacky style, because the sports, artworks, scientific breakthroughs and other subjects they depict were chosen by the company's employees and genuinely reflect their interests. By contrast, BP's green and yellow sunflower-inspired logo does not. BP adopted that emblem and dropped its old name, British Petroleum, in 2001 at a time when ecological protests against the oil industry were growing, and it wished to present itself as a responsible company that cared about the environment. Whenever BP has been embroiled in ecological disasters, such as the 2010 explosion in the Gulf of Mexico, its sunflower logo has been ridiculed as insincere and inappropriate.[33] Understandably so.

Structural integrity matters too. Anything that is unduly fragile or unreliable, like the perpetually jamming M-16 rifle, fails this test. As do show-stoppingly alluring chairs that are neither strong nor comfortable enough to be sat upon. They have so little integrity that they are not simply badly designed but 'ridiculous', as Donald Judd put it. Nor can design have integrity if, like those chairs, it is simply showing off. Take the gun lamps devised by Philippe Starck for the Italian lighting company Flos.[34] They are structurally sound and fit for purpose, in that they illuminate a room perfectly well, but modelling the base of a lamp on a Beretta or AK-47 was sensationalistic. Trading on the taboo of a potentially deadly weapon to sell a lamp is not an act of integrity. The same goes for skyscraping buildings, such as the Burj Khalifa in Dubai, whose architects were charged with making them as tall as possible. They might be popular with tourists who want to peer down from the top and city officials who enjoy bragging about having commissioned a record-breaking structure, and may even make striking additions to the cityscape, but they are too bombastic to be well designed. In *Design and Truth*, Robert Grudin levels the same accusation against St Peter's Basilica in Rome, whose elegant design by Donato Bramante and Michelangelo was destroyed by Pope Paul V's determination to enlarge it into the world's biggest church in the early 1600s as a monument to the power of the papacy.[35]

Critically, integrity also embraces environmental and ethical responsibility in design. Scientists may still be squabbling over the causes and eventual impact of the ecological crisis, but the situation is so grave that no one can pretend to ignore it. Nor can most of us claim to be unaware of the wider implications of the things we buy. If we have any reason to feel guilty or even uncomfortable about the ethical or environmental consequences of a design project – or the way it was conceived, developed, manufactured, shipped, sold and will eventually be disposed of – it cannot be considered to have integrity, and is therefore disqualified from being good design. How can we be expected to take pleasure in something that we suspect of being harmful either to the environment or to other people?

Take a seemingly mundane example: the espresso pod. Making an espresso with one of those neatly sealed little pods is undeniably faster, more efficient and less messy than doing so with ground coffee, and the result is likely to be more consistent. But even the most devoted pod-o-phile must baulk at all the packaging used by those tiny portions of coffee and the boxes they come in. Not only is there an awful lot of superfluous packaging, but some of it is not recyclable. In other words, the functional strengths of the espresso pod are negated by its environmental weaknesses and dearth of integrity. (Personally, I consider making espresso in the traditional way to be more pleasurable, and would cheerfully trade the pod's dependability for the delicious smell of ground coffee, and the frisson of waiting for it.)

Or consider the reassessment of the design merits of the world's best-known chair, the one that more people have seen and sat upon than any other. It is not the work of a historically important designer or a contemporary auction star, but the cheap chunk of plastic known as the monobloc, which has been manufactured as a single piece of polypropylene weighing two and a half kilograms since the 1980s.[36] No one knows how many have been made, but once you become aware of them you notice monoblocs everywhere: perching in the corners of car parks and construction sites; floating in the debris of typhoons, hurricanes and other natural disasters; and appearing in the background of TV news footage on alleged atrocities in military prisons, like Abu Ghraib in Iraq and the capture of terrorist cells and deposed dictators.

Unprepossessing though it looks, the monobloc has some design virtues. As well as being inexpensive, it is compact, portable, stackable, waterproof and easy to clean. In other words, it boasts many of the functional qualities that responsible designers prize, but it is also cursed by grave environmental flaws. If the polypropylene breaks, it is impossible to repair and, because it is not biodegradable, landfill sites are doomed to be stuffed with the corpses of unwanted monoblocs for decades to come. For the same reason as the espresso pod, the monobloc can no longer be deemed to be well designed.

Other lapses of design integrity occur when dubious environmental and ethical claims are made, as they were for a bunch of bananas sold in a London supermarket that were swathed in what was proudly described as 'organic packaging'.[37] Why waste material, regardless of whether it is recycled or recyclable, on a banana, whose skin is a perfect example of natural packaging that protects the fruit and can be easily removed when you want to eat it, before decomposing quickly and safely once you have finished? To do so is idiotic, but trying to score eco-points by billing it as 'organic' is dishonest.

The environmental case against those over-packaged bananas is unusually straightforward, just as it is for the espresso pod and the irreparable, unrecyclable monobloc. Assessing the ecological and ethical impact of most design projects is not: it is fraught with problems, and dissent is rife. One person's certainty – or even their idea of an acceptable compromise – is often another's bone of contention. Take the Toyota Prius, the first mass-produced hybrid vehicle to be powered by both a petrol engine and an electric battery.[38] The Prius consumes less fuel than most other cars, and that is its selling point. Some people feel not merely guiltless but virtuous when driving it, confident that they have made an environmentally responsible choice. Others disagree. They criticize Prius drivers for what they regard as their irresponsibility in choosing a hybrid vehicle when more energy-efficient cars are available. Then there is the ongoing debate about the Prius's battery, and the environmental implications of the materials it is made from. Should you feel guilty about driving a Prius? The answer will depend on whom you ask.

There is also the question of how design is used, the integrity of its context. Again, this can compromise something of otherwise irreproachable quality, which was the fate of two totems of twentieth-century design, the Egg and Series 7 chairs.[39] Originally designed by the Danish architect Arne Jacobsen in the 1950s for one of his buildings, the new SAS Royal Hotel in Copenhagen, they were then licensed for production by a local furniture manufacturer, Fritz Hansen. Both chairs are excellent examples of intelligent design. Their curvaceous forms are typical of the Scandinavian style of organic modernism, which was championed by Jacobsen as a gentler alternative to Le Corbusier and Mies van der Rohe's work. His chairs look unusual enough to be striking, yet not so much so that they ever seem obtrusive. Stylistically, they are also subtle enough to have proved surprisingly enduring: by appearing fashionably organic in the 1950s, 1970s and 1990s, yet futuristic in eras when that style was popular. Above all, they convey a sense of modern elegance, which is luxurious without being snooty, and imposing but not intimidating. One glance at them signals 'good design', even to someone who has never heard of Arne Jacobsen.

A few years ago, Fritz Hansen agreed to sell several thousand Eggs and Series 7s to McDonald's as the seating for its European fast-food restaurants. McDonald's customers were becoming fussier and the company was anxious to upgrade its image. To do so convincingly, it needed to make its restaurants more appealing. What better way of doing so than to introduce Arne Jacobsen's impeccably designed chairs? It could be argued that this was a good thing. After all, the more people who have the opportunity to experience good design the better, regardless of whether they are perching on it to scoff a Big Mac or Filet-O-Fish. But selling those chairs to McDonald's was risky for Fritz Hansen. Would they still be associated with modern luxury, or with Ronald McDonald, McMuffins and McNuggets?

But McDonald's did not stop there. Having installed the original Fritz Hansen chairs, it decided to supplement them with less expensive replicas in countries like Britain, where the original designs were out of copyright.[40] McDonald's then commissioned new chairs, supposedly 'inspired' by Jacobsen's designs, which looked like parodies of the originals. Fritz Hansen refused to supply McDonald's again. By using Jacobsen's chairs in such a hostile context, McDonald's had deprived them of the very qualities it had paid so much money for, and turned a textbook example of good design into a cautionary tale of the danger of forfeiting design integrity.

4 Why good design matters

Accidents, disasters, crises. When systems fail we become temporarily conscious of the extraordinary force and power of design, and the effects that it generates. Every accident provides a brief moment of awareness of real life, what is actually happening, and our dependence on the underlying systems of design.
— Bruce Mau and the Institute Without Boundaries[1]

The official alarm was sounded on 28 February 2003 when the French Hospital of Hanoi notified the local office of the World Health Organization about a patient who had been admitted with an unusually aggressive influenza-like virus. He was Johnny Chen, an American businessman who lived in Shanghai and had flown to Hanoi from Hong Kong two days before. On arrival he was taken to the hospital, a tiny private clinic, whose staff were concerned that his strange illness was a form of avian influenza virus, or 'bird flu'. The WHO dispatched Dr Carlo Urbani, a specialist in infectious diseases, to investigate. He was so alarmed that he stayed at the hospital for the next few days, assisting the staff, enforcing infection controls and documenting Chen's condition.

Based on his recommendations, the Vietnamese government quarantined the French Hospital on 9 March and imposed other emergency measures in a desperate effort to contain the disease. The WHO then issued a global alert about an outbreak of severe acute respiratory syndrome – SARS for short – an aggressive, highly infectious, often deadly disease, which had emerged in China the previous autumn and was spreading with terrifying speed throughout Asia. Within weeks of Chen's death, SARS had been detected in thirty-seven countries and was known to have killed nearly a thousand people, including Dr Urbani and other members of the medical team who had treated him in Hanoi.[2]

At the height of the SARS alert, a senior designer for an American company was scheduled to fly to China to oversee the final stage of prototyping for a new product in a subcontractor's factory. He made such visits several times a year, generally staying in China for a week or two. This visit was particularly important because the product, the result of years of research and development, was expected to be a best-seller and the company's financial forecasts had been set accordingly. But there was a problem. Americans, like other nationalities, had been advised not to travel to Asia for fear of contracting SARS. The company's insurers refused to provide cover for the designer or his colleagues to go there, and it proved impossible to find an alternative source of insurance. Unless he travelled to China to complete the design process, the new product would not be completed on time. Nor was there any indication of when the SARS alert would end. No one knew how long it would take to prevent a pandemic by controlling the spread of the disease, or whether it would be possible to do so.

The company was so desperate to finish the project on schedule that it chartered a long-haul aircraft and kitted it out as a home-cum-design studio where the designer could live and work. He then flew to China in the plane, which was 'parked' on a runway of the airport closest to the factory. A controlled system was set up to deliver prototypes, components, tools, testing equipment and anything else he might need. To prevent anyone inside the aircraft from risking exposure to infection, they arrived there in surgically sealed containers. The designer assessed the new product, discussed his findings and suggested modifications with the prototyping team at the factory using a video-conferencing program. He stayed on the aeroplane for ten days without ever setting foot outside it, then flew back to the United States.

Why did he do it? Professional pride? Corporate loyalty? Determination to complete a project on time, having devoted several years of his working life to its development? Sheer bloody-mindedness? Any or all of those factors may have prompted his decision to cut himself off from the world in the safe but far from pleasant environment of that aircraft for so long. It is easier to understand why his employer should have invested so much effort and expense in ensuring that the design of its new product would be up to scratch and finished on deadline. Money. The company stood to make much more of it if the design project was completed on time, and considerably less if it was not.

Money. Money. Money. It may not be a fashionable or romantic explanation for why good design matters, but it is undeniably persuasive, and has been throughout history. Yet design matters in many other respects too. How could it not, given its ubiquity? Design exerts so much power over so many aspects of our lives that the quality of a design project can be a decisive factor in deciding whether we will enjoy happiness and success, or be subjected to misery, failure or worse.

The design historian John Heskett once likened design to language, arguing that each is 'a defining characteristic of what it is to be human'.[3] Like language, design is unavoidable, and our relationship to it – the degree to which we understand what is being communicated to us and to which we can express what we think, feel and desire – has an immense influence in determining how we deal with the world, and it with us. There are similar parallels between design and health. Regardless of how much or how little we choose to dwell on it, whether our health is good or bad has a huge impact on our quality of life. A sense of well-being can be empowering and nurturing, while poor health can be enfeebling, painful and possibly fatal. Even people blessed with strong constitutions and good genes may have to deal with some sort of medical difficulty at some point in their lives. Few of us can escape this fate, but there is much we can do to mitigate the damage, by spotting potential problems swiftly, taking preventative action by eating sensibly or exercising, and choosing the correct medical treatment. Exactly the same can be said of design.

Consider the impact of good design on an individual: on Aimee Mullins, the American actor, model and athlete who set three world records for the hundred metres, two hundred metres and long jump at the 1996 Paralympic Games in Atlanta, Georgia.[4] She is a bilateral amputee who was born without fibulas in both legs, which were amputated below the knees on her first birthday. Her parents were told that if she kept her lower legs she would have to use a wheelchair for the rest of her life, but if they were amputated she could learn how to walk with prostheses. The problem was that for anyone growing up in the United States in the late 1970s and 1980s as Mullins did, the only available artificial limbs were woefully badly designed, as they had been for decades.

'My earliest prosthetics were little more than rudimentary stilts,' she recalled. 'They were made from a wood-plastic compound material with rivets on the sides of the knees and rubberized feet held on to the shins by metal bolts. There was no give when you walked, and your entire body weight landed in just a few places on your residual limb, which was often times why it blistered.' Equally painful were the leather straps with which she attached the prostheses to her thighs. 'I had to tie them so tightly that they killed the circulation in a death-grip strangling effect. The residual limb ended up being desensitized in some places and highly sensitized in others, which is why some amputees opt to use wheelchairs, because they can't adapt to the initial levels of pain.'[5]

Those prostheses were also hopelessly unsuitable for a sporty, sociable child like Mullins, who grew up in Allentown, a small industrial city in Pennsylvania. The 'toes' were prone to breaking if she kicked a ball or scuffed them on a diving board, and swimming in her prosthetics was strictly forbidden in case the wood rotted and the bolts rusted.[6] 'But I was a child, and on a hot July day there was no way I was staying out of the water.' Eventually she was given a pair of waterproof polypropylene legs, which were so buoyant that she would dive into the water only for them to drag her straight back up to the surface, until her father drilled holes in the ankles. 'Every pair of legs I had we ended up hacking somehow in my dad's tool shed.' Worst of all was when the wood broke. 'I'd taken a ball in the shin during a game. Of course, it didn't hurt at the time, and we didn't realize that it had formed a hairline crack in the wood. Six months later, I was jumping around in a music class – we were doing the twist. There was this horrible splintering sound, then kids screaming and the teacher fainting at the piano. All I remember thinking was that my parents were going to kill me, because getting a leg on the insurance was such a nightmare.'

When Mullins was sixteen, her wooden legs were replaced by woven carbon-fibre ones, which were, at least, less painful and onerous to move in. 'The first time I put them on, I felt as though I was walking on a cloud. I hadn't realized what an ordeal it had been to wear the wooden legs, because they were all I'd ever known.'[7] Not that her new prostheses were entirely painless, or effortless; and they looked dreadful. 'The quote-unquote cosmetic covering was made from a horrible dense foam. And they were unisex, with two colour options: "Caucasian" and "Not". "Caucasian" was the ugliest shade of peach you have ever seen. Growing up, I didn't meet another amputee until I was a teenager, and I didn't think of them as being my tribe, any more than blondes. I understood that I was different, but other people had differences too. Putting on my legs seemed no different from a friend putting in contact lenses. But I do remember looking at a waxwork of Jerry Hall at Madame Tussauds on a trip to London, and thinking: 'If you

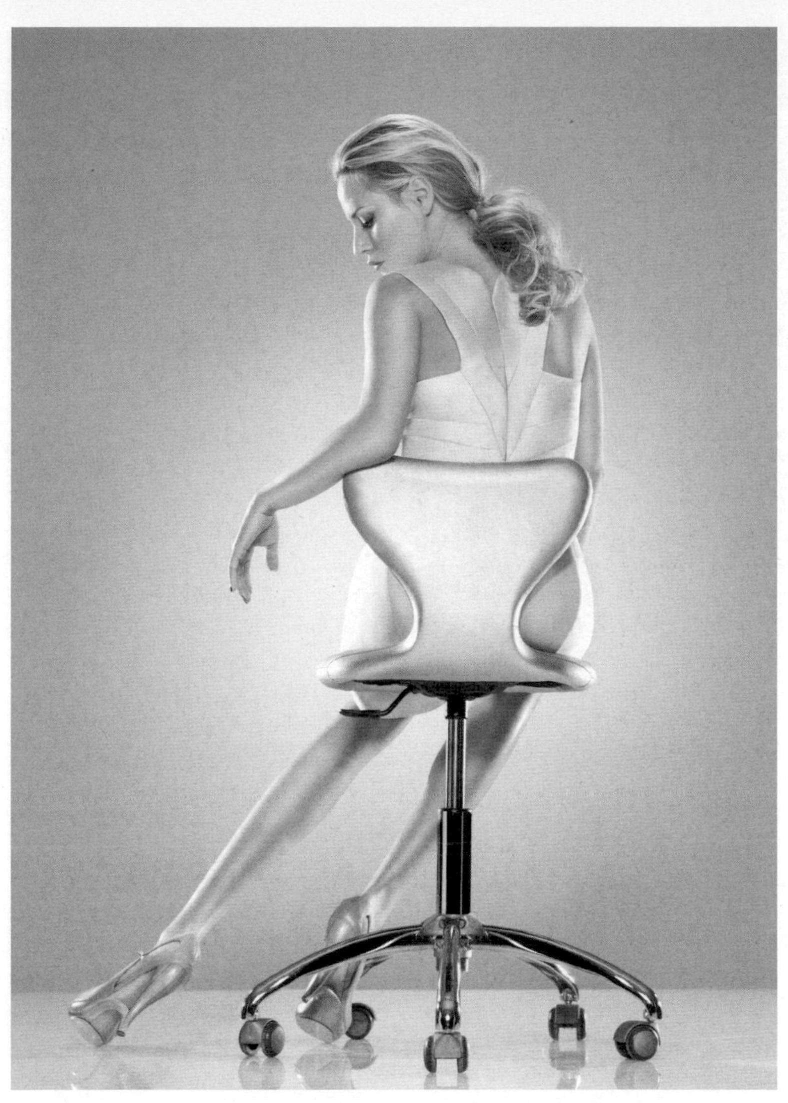

Aimee Mullins wearing silicone legs

Aimee Mullins wearing fantastical bespoke prosthetic legs in Matthew Barney's 2002 work *Cremaster 3*

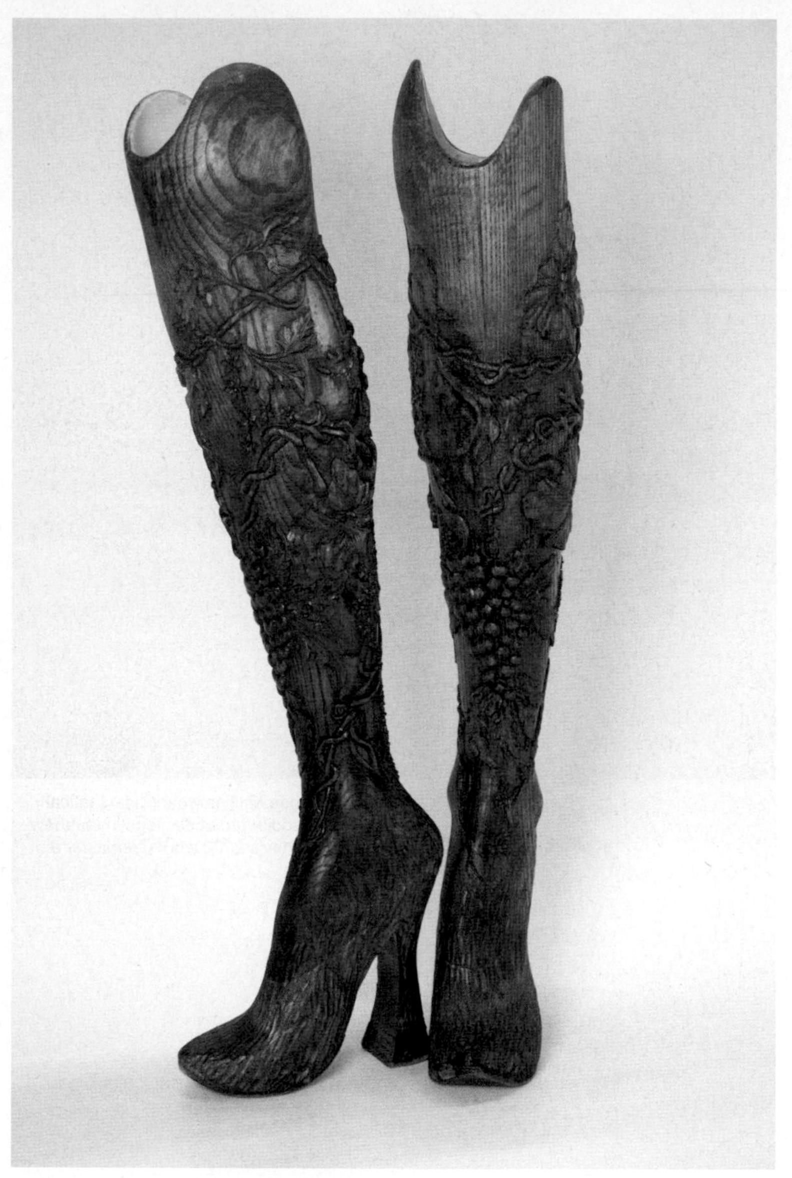

Bespoke wooden legs worn by Aimee Mullins
in an Alexander McQueen fashion show in 1999

can build this mannequin, with the layering of texture to propose tendons and muscles, and the specificity of colour and shape, why can't you build a decent prosthetic for human beings?'[8]

Having excelled at sports in high school, Mullins participated in track and field events as a student at Georgetown University in Washington, becoming the first ever amputee to compete in US collegiate sports. It was then that she began experimenting with the design of her prostheses. She started in 1995 by being the first person to be fitted with the woven carbon-fibre sprinting limbs, which are modelled on the hind legs of a cheetah and are now standard issue for athletes wearing prosthetic legs. Mullins competed in them in the 1996 Paralympics. 'They were incredible, but off-track I still had to put on these horrible foam-covered legs,' she said. 'They were made for dead-flat orthopaedic shoes, but I'd shove the foot into a shoe with an inch of a heel. I was a teenager and I'd put up with the pain of being pitched slightly forward, straining my hips and knees, to look cool. I remember someone saying: "It's really a shame that you care what you look like, Aimee. You're an amputee, you just need to accept that." To me, that statement about "acceptance" was really shocking. I wasn't going to be embarrassed or ashamed for demanding more than function from prosthetics.'

Since then, Mullins has sought to overcome the design deficiencies of her prostheses by working with prosthetists, biomechatronic engineers, athletes, artists and designers to develop the type of legs that meet her needs and expectations. She now has fourteen pairs, nine of which she uses regularly. For everyday use at home in New York, she wears woven carbon-fibre 'shock absorber' prosthetics, which are comfortable and practical. 'If I wear them for running around the park or going to the food market, children come up and ask: "Can you fly?" Or they'll say: "You should put rocket boosters right there."'[9]

Other legs were designed specifically to look like human flesh-and-bone legs. Since 1997, Mullins has had silicone legs made for her by the prosthetist Bob Watts at Dorset Orthopaedic in southern England.[10] Remarkably realistic to look at, each leg can be made in whichever shape and length she chooses, and the feet are angled to fit the different heel heights of her shoes, going as high as four inches. To make them as lifelike as possible, Watts has added tiny flaws, such as veins and birthmarks, to the silicone 'flesh', though Mullins drew the line when he suggested ageing a new set of legs by adding more veins. 'I was like: "Oh no, no, no!"'[11]

She has also developed fantastical legs. When the fashion designer Alexander McQueen invited her to model in one of his shows, they collaborated on the design of a pair of wooden legs, which were carved from solid ash by a master craftsman into ornate and clearly artificial prostheses.[12] She worked with the artist Matthew Barney to devise a surreal collection of limbs for her role in his film *Cremaster 3*, including transparent glass legs and replicas of the sleek, furry hind legs of a cheetah that transformed Mullins into a mystical creature.[13] 'I don't want to promote the idea that human-looking legs are more desirable than non-human-looking legs. It's a throwback to amputees feeling that they need to "pass". I quite like my carbon legs and feel that they are desirable as an aesthetic choice as well. I want people to find empowerment from personal choice among many different aesthetic options. If you have so many choices when it comes to a couch or an iPad, why not have the same choices with something as intimate as a prosthetic?'

Rather than risk being constrained by the badly designed prostheses she was originally given, Mullins has chosen to make herself faster, stronger, taller and weirder by commissioning specific types of legs with which she can enhance her natural beauty, athleticism and theatricality. One night, when she arrived at a party in New York wearing her longest, most elegant prosthetics with her highest heeled shoes, a friend came up and said mock-seriously: 'Oh Aimee! That's not fair.'[14]

In future, Mullins expects her prostheses to become increasingly efficient, versatile and appealing as technology advances. An important area for innovation is joints, and a team of researchers at the Massachusetts Institute of Technology, led by the American engineer Hugh Herr, head of the Biomechatronics Group, has pioneered the development of a powered ankle-foot, which replicates the movement of a biological ankle.[15] Mullins's first set of powered ankles were made by him. 'They save my knees and hips from having to withstand so much impact, and enable my muscular system to work more efficiently.'[16]

Like her, Herr is a double amputee. An avid mountaineer, he lost both his legs below the knee in a teenage climbing accident, when he and a friend were trapped in a blizzard in an icy ravine in New Hampshire. He, too, struggled with inadequate prostheses, and has devoted his work in biomechatronics to the design of more efficient artificial limbs. Before developing the powered ankle, he made a similar breakthrough with the knee, but has also developed bespoke legs for himself. After his accident, Herr was determined not only to climb again, but to continue to participate in free-climbing competitions, and he constructed a pair of prosthetic legs designed specifically to enable him to do so.

Free climbers forgo ropes, harnesses and other protective equipment, making the sport so dangerous that a climber's safety relies entirely on his or her physical strength and agility, climbing ability and psychological fortitude, which, in principle, could put a bilateral amputee at a disadvantage. Before his accident, Herr was an accomplished free climber, and he designed his new prostheses to amplify his natural prowess. His bespoke legs can be lengthened or shortened to the precise length needed for him to hoist himself up – or down – to the next level of a climb. If necessary, he can vary the length of one limb from the other. Another advantage is that the tips of his prostheses are so much smaller than human feet that Herr can find secure footholds in the narrowest nooks and cracks, which would otherwise be too tiny to cling on to. So effective are his artificial climbing limbs that some of the fellow free climbers who had pitied him after his accident were soon calling for him to be banned from entering competitions because of his 'unfair advantage'.[17]

Those ingeniously designed prostheses are compelling examples of how much good design can matter to individuals, in this instance to people who were determined, resourceful and imaginative enough to take matters into their own hands. Having identified the design flaws of critically important components of their lives – the artificial limbs they needed to walk, run, swim, jump, dance and climb – Aimee Mullins and Hugh Herr took the action required to develop superior alternatives. Millions more people have benefited from their efforts, which have contributed to general improvements in the design of artificial limbs all over the world. Their achievements demanded courage and resilience, but they did at least have the advantage of being able to identify the cause of the problem – the shortcomings of the design of their original prostheses. It is much more common for design to affect our lives without us noticing that it has done so, which can make it even harder for us to address the problems caused by dysfunctional design.

Think of the World Cup soccer tournament. It is the world's most popular sports event, watched by more people than any other, and its emotional impact is incalculable. Psychologists have written book after book on why people feel able to express the joy, exhilaration, despondency, fury and other emotional extremes during soccer matches that they repress in more portentous situations, such as falling in love or faltering in their careers. During the World Cup, those emotional extremes are heightened because national reputations can be won or lost, old loyalties put to the test and new allegiances formed.

The underlying principle of the World Cup, and all other sporting events, is the importance of fair play. With so much at stake, it is essential that every team has an equal chance of success and each player is treated in the same way. The primary means of enforcing these principles are the rules of the tournament and the conduct of the referees, but the design of various elements of the sport are relevant too, including the new ball which is developed for each World Cup. The Fédération Internationale de Football Association, or FIFA, the Zurich-based body that organizes the tournament, commissions a new ball every four years from the German company Adidas, which pays a hefty fee for the rights to be the official supplier. Adidas does so in the knowledge that its newly developed ball will be seen on television by hundreds of millions of people, many of whom will buy it.

In the interest of fair play, the World Cup ball must behave in exactly the same way wherever and whenever it is kicked, regardless of the conditions, thereby ensuring that no team has an unfair advantage over another. What could possibly go wrong? How hard can it be to design a round object of a specific size and shape? Designers know that the smoother and rounder a soccer ball is made, the likelier it will be to respond in the same way to the impact of a player's foot. They also know that it should absorb as little moisture as possible during the course of each game to avoid fluctuations in weight. But identifying what sort of ball to design is the easy part. Producing it is more difficult thanks to the complex physics of the sphere, which scientists still know less about than the aerodynamics of aircraft and Formula 1 racing cars.[18]

For decades, most professional soccer balls, World Cup balls included, adhered to the late nineteenth-century design template of eighteen stitched leather panels. During the 1966 World Cup in England, the growing number of television viewers complained that they could not see the ball clearly enough to follow the games on their black and white TV sets. It was then that FIFA charged Adidas with designing a telegenic alternative. The result was the Telstar, which was made from twelve black pentagonal panels and twenty white hexagonal ones. It was introduced at the 1970 World Cup in Mexico, when it proved hugely popular, not least because it was clearly visible on television, and has been the default design for soccer balls ever since. The Telstar is a glowing example of design's ability to enhance our lives, both by solving a practical problem and improving our sense of well-being. By enabling many millions of soccer fans all over the world to follow the progress of World Cup games clearly, it has given them great pleasure (at least when their teams have won) and defused the risk of their friends and families being subjected to furious complaints about the lousy picture quality on their television screens.

Adidas has continued to produce a new World Cup ball every four years, but with mixed results. The Tango España ball was a disaster at the 1982 tournament in Spain. The rubber inlay over the seams rubbed off when it was kicked, and the ball had to be replaced during some games, disrupting the flow of play and irritating everyone: players, managers, officials, spectators in the stadiums and television viewers alike. Whereas the Questra ball shone at the 1994 World Cup in the United States. A newly developed polyurethane foam coating made it faster in flight, arguably helping the players to score more goals.[19] The +Teamgeist ball at the 2006 World Cup in Germany had a similar effect. It was made from fourteen panels bonded together by thermal technology rather than by traditional stitching, to create a smoother, water-resistant surface. Strikers loved it, as they could kick the ball more powerfully from long distances. When Germany played Costa Rica in the opening game of the tournament, the German midfielder Torsten Frings scored a spectacular goal by kicking the ball thirty-five yards into the goalmouth. It started off straight then bent sharply to the right for the last ten yards.[20] But goalkeepers were less enthusiastic about the dynamic new ball. One of their complaints was that it moved erratically because of its smoothness. Another was that the ball was so light it often slipped out of their hands. Though, like the Questra, the +Teamgeist did make World Cup games faster and more exciting.[21]

The design of the Tango España, Questra and +Teamgeist undoubtedly made a difference to their respective World Cups, but did not necessarily detract from the principle of fair play, because each player was equally likely to be advantaged or disadvantaged by their idiosyncrasies. The Jabulani, the official ball for the 2010 World Cup in South Africa, proved more controversial. No sooner had the tournament begun, than goalies started to complain of it swerving, slipping and spinning erratically. The problem lay not with the ball itself, but with its response to the stark differences in altitude and air density between the South African cities where the games were played, particularly between Johannesburg and Cape Town. The ball was likely to swerve less and to fly to greater heights in the former, which is at a higher altitude than the latter.

Another anomaly was that the ball's position could change by anything up to two diameters in a typical goal shot between the two cities, making it difficult for goalies to anticipate its path, especially as teams often flew straight from one place to another for successive games. When that happened, the goalkeepers' chances of guessing where the ball would land lessened considerably.[22] In theory, this meant that a team which had switched altitude between games was at a significant disadvantage to one that had not, and the failure to produce a ball which was immune to such changes had compromised fair play.

Not only do the fortunes and misfortunes of the various World Cup balls illustrate how much design can matter to us without us even realizing (in this instance by possibly distorting the outcome of an event that many millions of people care passionately about), they also show how much likelier we are to notice design's failures than its triumphs. How many of the strikers whose chances of scoring were enhanced by the speed of the Questra or power of the +Teamgeist credited their goals to those balls? None. Yet goalkeepers wasted no time in grumbling that the +Teamgeist was slippery and the Jabulani erratic. Perish the thought that those goalies might have dropped the ball after bungling a catch; or that a triumphant scorer would admit that, if not for its quirks, the ball might not have ended up inside the goal.

When design affects us directly, rather than being something we see from across a sports stadium or on television, the same principle applies. Many of the greatest design feats go unnoticed: they do their job so well that we are barely aware of them, happy to congratulate ourselves for having solved a problem or made some sort of progress. One of the most important functions of design is to regulate our behaviour, whether it is by pointing us in the right direction, enabling us to operate an otherwise impenetrable digital device, or saving us from danger. If the end result is well designed, the experience of using it will be so uncomplicated and instinctive that we will not need to think about it.

Take Zurich Airport. For years, whenever I went there, I was struck by how easy it was to find my way around and how calm it felt, just as airports are supposed to feel but rarely do. It was not because of the architecture. The glass and steel sheds look exactly like those of hundreds of other airports. Nor was it the location, which is indistinguishable too. It was thanks to the airport signs.

There was nothing especially stylish or noticeable about those signs, but they always seemed easy to spot and to decipher. The letters and numbers were presented in a crisp, clear typeface. The illustrated symbols were instantly recognizable black silhouettes on white squares: a knife, fork and spoon for the café, and so on. Unlike the blizzard of signage that greets you at other international airports, such as Heathrow in London or JFK in New York, there were not very many signs in the terminal buildings at Zurich, but whenever I needed to check where I should be going, one always seemed to appear.

Those signs were the work of one man – the Swiss graphic designer Ruedi Rüegg – for nearly forty years.[23] He designed other signage schemes, for the Zurich Opera House and nearby Basel Airport, but Zurich Airport was his masterpiece.[24] Rüegg updated the signs there regularly, adapting them to changes in technology or the airport's layout, but the colours, symbols and typefaces stayed the same. The original system was so thoughtfully designed by him in 1972 that there was no need to alter it, at least not dramatically.

Rüegg's design scheme guided me through Zurich Airport so deftly that being there always felt soothing to me and, I suspect, to other people too. But how many of them would have noticed those intelligently conceived and positioned signs? Very few. Yet design projects like this one are shining testimonies to the value of good design and its ability to make a positive difference to our lives. Rüegg's signs spared people the anger and frustration of missing a turning, walking too far in the wrong direction or, worse still, arriving too late for a flight. Even so, it is not surprising that we often fail to recognize the value of something that saves us from making mistakes; it is only when design projects like this one go wrong that we realize quite how much they matter.

How many times have you worried about losing your way in an airport because of poorly designed signs? Missed a train thanks to an indecipherable timetable? Taken a wrong turning in your car after struggling to decipher a confusing road sign? Thrown away one form and filled in another because the layout was so befuddling that you were not sure which boxes to fill in and what to write there? Or given up trying to do something with your computer because, however many keys you press, you do not seem to be able to work out how to do it? By demonstrating how damaging bad design can be, such glitches remind us of how dependent we are on good design. If the signs, timetable and form had been clearer, and your computer easier to operate, would you have noticed? Probably not. They would simply have fulfilled their functions efficiently, rather than standing out as stellar examples of design. But bad design is unavoidable, because its effects can be so grave, even if the cause of the problem seems inconsequential.

This was the case with the apparently minor flaws in the design of the ballot cards for the 2000 US presidential election in Palm Beach County, Florida.[25] When the local voters arrived at their polling stations on 7 November 2000, they were given punch cards listing all ten presidential candidates, including the front-runners, Vice President Al Gore representing the Democrats and George W. Bush for the Republicans. The cards were inserted into a marking machine so that each voter could punch a hole in the perforated square next to the name of their chosen contender. If all went well, a small square of paper, known as a 'chad', would be removed beside the correct candidate's name. But all did not go well that election day.

One obstacle was that some of the Palm Beach County marking machines turned out to be faulty, and failed to dislodge all the chads. The holes in some cards were partially perforated and others not at all, raising the risk of them being rejected as 'spoilt' ballot papers by the electronic equipment that counted the votes. This problem was down to mechanical failure, but another glitch concerned the design of the punch cards. In the preparations for the election, Theresa LePore, the official responsible for organizing the voting in Palm Beach County, had been concerned that if all ten candidates were listed on the same page, as they were everywhere else in Florida and in many other counties across the United States, the typeface would be too small for elderly voters to read: a common cause of complaints about ballot card design. Her solution was to spread the names across two facing pages, in what is called a 'butterfly ballot', thereby allowing the printers to use bigger, clearer type.

LePore had acted with the best of intentions.[26] At the time, the changes she implemented may very well have sounded sensible to her colleagues and the Palm Beach County electorate. But on polling day, a worryingly large number of Democrats complained of having been so confused by the new layout that they had mistakenly punched a hole beside the name of the ultra-conservative Reform Party candidate Patrick J. Buchanan, thinking that they were voting for Al Gore. It is not surprising that they did so. Gore's name, the second to be listed on the left side of the ballot paper, was almost directly across from Buchanan's on the right, and the holes for voters to punch were side by side in the centre. Other voters claimed that they had panicked and punched more than one hole, thereby invalidating their ballot papers.[27] In other words, the final tally was bound to have been distorted because one batch of votes would have been recorded for the wrong candidate, and another batch rejected.

It goes without saying that, in the interests of fairness and democracy, every election result must be as precise as possible, but accuracy becomes even more important when the outcome is as tight as it was in the 2000 US presidential poll, both nationally and in Florida. The voting there turned out to be so close that a recount was announced the following day. Under Florida law, a recount is compulsory whenever an election is decided by a majority of less than half of one per cent of the total vote, as it was in the first count, in which George W. Bush beat Al Gore by just one thousand, seven hundred and eighty-four of the six million votes cast. Recounts are always dramatic in such important elections as presidential polls, but this one was especially so: the national vote was so tight that whichever of the front-runners won Florida would win the presidency.[28]

As the recount was manual, it should have rectified at least some of the mistakes made by Palm Beach County's faulty marking machines. But how could it have determined which of Pat Buchanan's 'supporters' had really intended to vote for Al Gore? Or which of the multiple holes in a 'spoilt' ballot paper was the correct one? It is impossible to judge exactly how many votes were credited to the wrong candidates in Palm Beach County because of those badly designed punch cards, and how many were simply 'lost'.[29]

What we do know is that Palm Beach County is a Democrat stronghold, yet Pat Buchanan (whose politics are so right-wing that Wikipedia describes him as a 'paleoconservative') won three thousand, seven hundred and four votes there,[30] three times more than he received in any other Florida county, even the conservative strongholds. And he had not made a single campaign stop in Palm Beach County, presumably because he and his advisors had not considered it to be worthwhile. Local

Democratic officials claimed that Al Gore lost as many as three thousand votes to Buchanan because of the confusingly designed layout, which would have been more than enough for him to have defeated George W. Bush and clinched the presidency.

Would Al Gore have become the forty-third president of the United States of America rather than George W. Bush if Palm Beach County's voters had been given a single-page ballot card designed like those in all of the other Florida counties when they went to the polls on 7 November 2000? Possibly. And would history have been different if Al Gore had won? Undoubtedly. That is how much good design can matter.

5 So why is so much design so bad?

I have never been able to understand why cheap books should not also be well designed, for good design is no more expensive than bad.
— Allen Lane[1]

If there is a designer who begins his or her working day determined to design something that is bad, or maybe just mediocre, I have yet to meet them.

At its best, design is a noble endeavour, committed to making the world a better place. Not that every designer is talented or lucky enough to work on such heroic projects as developing an energy-efficient car or educational software that could help the world's poorest children to learn how to read. The dull truth is that they are likelier to end up dedicating years of study and cherished dreams to humdrum tasks like redesigning toothpaste packets, or cutting the cost of producing a particular model of washing machine because its manufacturer has been acquired by a private equity group with usuriously high profit targets. Even so, I do not believe that any of them wants the outcome to be disappointing.

So why is it? If you doubt that most design is bad, just look at the evidence on a retail website or in a shopping centre. Many, though thankfully not all, of the things you see there will have few if any of the qualities we associate with good design. They are unlikely to be particularly useful, desirable or responsible, and if any of those virtues are evident, they may well be countered by the absence of others. Whereas they will very probably be ugly, inefficient, derivative or wasteful. Neurotically over-styled cars. Illogical signage systems. Illegible typography. Inscrutable instruction manuals. Pretentious corporate symbolism. Anything that guzzles energy unnecessarily or is impossible to recycle safely. Examples of bad design are frighteningly common, much more so than good ones, and there are many more instances when design is not necessarily bad, but so uninspired that it is not good enough.

You could argue that design is no different from anything else and that every other area of life suffers the same problem. How many artists can match Gerhard Richter's virtuosity, David Hammons's eloquence, Ai Weiwei's courage, Isa Genzken's acuity or Rosemarie Trockel's eclecticism? Very few. But that is no excuse. Art history tends to forget the disappointments and celebrate the triumphs, and the same tendency is apparent in design. But if we accept that no designer sets out to produce something substandard, and that the same can be said of the people they work with, why do they end up doing so?

Consider a car. It always strikes me as odd that an object which is the single most expensive thing that many people buy, except for their homes, should so often be badly designed. Ungainly exteriors, garish interiors, uncomfortable seats, irritatingly over-complicated dashboards and feeble ecological claims are just a few of its routine failings. What a dismal decline for the industry that pioneered a radical new system of mass production with the 1908 Ford Model T, and produced a car so gorgeous in the 1955 Citroën DS 19[2] that Roland Barthes nicknamed it *la déesse*, the French word for goddess.[3] If we drivers were to receive a fair return for our emotional and financial capital, a car should be one of the most desirable things we own. And it is not that there is a dearth of investment in new models. Quite the opposite. The automotive industry ploughs hundreds of millions of dollars into research and development every year in the hope of tempting us to buy new cars. So why are so many of us forced to compromise in our choice of vehicle? And why, despite those generous development budgets, has the industry failed to produce an energy-efficient equivalent of a vintage beauty like the DS 19?

One obvious explanation is that designing anything well is tough, particularly if it is as structurally challenging as a car. Take one element of its design – the bodywork. The designers need to ensure that it is strong enough to support the engine and suspension system, as well as to hold up the roof. Having satisfied those functional requirements, they then need to refine the form. There are sensitive intersections of surfaces to deal with, often in different materials. An additional complication is that the vehicle's shape is generally conceived as a series of optical illusions. If a line looks straight on a car, it will actually be a curve or one of the irregularly curved lines that mathematicians call 'splines', which were designed to seem straight. All of the 'curves' are probably splines too, and any parts of the bodywork which appear symmetrical will be subtly different to create the illusion of similarity. Pulling that off is tough.

The end result also has to conform, quite rightly, to a maelstrom of health and safety regulations, as must every other component of the vehicle. Automotive designers often note ruefully that many of their own favourite examples of car design pre-dated the safety crackdown, including the DS 19 and the Lamborghini Miura, which was developed by the Italian designer Marcello Gandini with his colleagues in Lamborghini's engineering team during the mid 1960s, initially as a labour of love that they worked on in their spare time.[4]

Designers in every field face their own tensions and complications. Remember how refined the design of the typeface Helvetica is compared to Arial's? And how precise the functional requirements of Aimee Mullins's sprinting legs and Hugh Herr's free-climbing limbs need to be? However daunting those design challenges are, it is the designer's job to tackle them, which is why technical complexity is more of an excuse than an explanation for the substandard design of cars or anything else.

One cause of sloppy design lies in design culture, and with the type of people who become professional designers. Thankfully, there has been progress since Le Corbusier ejected Charlotte Perriand from his studio and Charles Harrison was told that the Sears design team did not hire African Americans.[5] But design is still a Western-centric discipline, and the most prominent designers are mainly male and mostly white. We need the best possible designers, and we will not get them unless the design community reflects society as a whole. Until it does, the quality of design will continue to suffer.

Design's lack of diversity contributes to another problem: the self-referential tendency of many professional designers. Too many of them suffer from 'designing for other designers' syndrome, whereby they seem to be striving to impress each other with their work, rather than the people who will use it. This is endemic throughout design, but particularly severe in fields like automotive design and fashion, where students enrol on specialist courses, often isolated from their peers, and work with fellow specialists after graduation.

Car designers tend to speak in jargon, which is incomprehensible to everyone else. They even draw in a similar style, depicting vehicles in sweeping lines with exaggerated proportions and heavily stylized details, such as unrealistically large wheels. Spending their university education and working lives surrounded by fellow specialists makes them less likely to absorb the broader developments that influence other areas of design and their customers' taste. Similar problems apply to fashion designers, who frequently end up living and working in a closed circle of stylists, photographers and editors. How many times have you looked at a fashion collection and wondered why nothing seems to be wearable? Or watched a woman hobbling in pain at the end of a party because the heels of her shoes are agonizingly high? Or hobbled yourself? I have, usually because I was silly enough to buy a pair of shoes which were designed to appeal to a fellow designer or stylist rather than to be walked in. Though thankfully I have never been quite as foolish as one French fashion editor who is said to have resorted to painkillers in order to subject her feet to freakishly high heels.

Another problem is that some areas of design tend to be prized above others, often to the detriment of the end result. The design of both cars and computers, for instance, has long been dominated by engineering. Understandably so. The single most important element of a car is its performance, and the emphasis on it has produced admirable advances in speed, safety and reliability. But problems arise when the focus on engineering detracts from other aspects of the design, such as aesthetics and usability. Similarly, computers must be reliable, but is it really necessary to encase so many of them in indistinguishable plastic boxes? And why are they so tricky to operate? Possibly because their designers and manufacturers were so tantalized by the dazzling new features dreamt up by the engineers that they expected their customers to be seduced by them too, even though their own research will have explained that very few of us ever use our computers to their full capacity.

The consequences can be even more damaging for products for people with disabilities, whose designers have traditionally come from engineering or clinical backgrounds.[6] Speak to anyone who uses a wheelchair, a hearing aid, or prostheses, and you are likely to hear a litany of complaints about their shortcomings, often because the human aspects of their design have been neglected. Even minor design gaffes can be infuriating.[7] The sociologist Tom Shakespeare has achondroplasia, a genetic condition which restricts leg and arm growth. He has a long list of ineffectual 'adapted products' that he has had to deal with over the years, and was particularly irritated to discover that a genuinely useful device – a pair of metal jaws mounted on a rod with a trigger action, which enables him to pick things up off the floor or from high shelves – had been patronizingly named 'The Helping Hand'. (Shakespeare has rechristened his 'The Grabber'.[8])

The way that designers work can be problematic too, particularly in a commercial context. Typically, they are organized in teams, which should, in theory, amount to more than the sum of their parts by enabling each member to play to their strengths, while colleagues compensate for their weaknesses. The risk is that the design process can become so fragmented, with different designers working on different parts of the same project, that the outcome may seem incoherent. Website design is particularly problematic, because navigation is often entrusted to 'user experience' specialists, aesthetics to graphic designers, the structure of the site to developers, and so on.

Most commercial design exercises are also subject to negotiation with members of other teams, such as marketing, finance, sales and engineering, who have a vested interest in the outcome but may have little enthusiasm for, or understanding of, design. Typically each decision is critiqued by committee, which inevitably stacks the odds against originality or innovation. The outcome may then be submitted to consumer focus groups, thereby raising the odds again. The investment required to develop a new product, or to modify elements of an old one, can be so high that making the wrong choice can be costly, financially and reputationally. No wonder corporate design culture is often conservative, and so many products end up looking pretty much the same.

Gifted designers can overcome these obstacles, but only if they combine raw talent with the necessary diplomatic skills and strength of will to steer their ideas through corporate bureaucracy, resisting the pressure to compromise. Many designers do not have the chance even to try, particularly if they work for external consultancies, which can have little or no control over what happens to their proposals once they have sold them to the client. Even the most powerful in-house designers can find the process tough. Dieter Rams, the celebrated head of design at Braun in its late twentieth-century heyday, once said that 'To be a designer, you have to be half-psychologist.' Though he also claimed that his job became a little easier once he realized he could win over the engineering team by buying 'a bottle of good cognac to share with them'.[9]

All designers, no matter how talented, diplomatic or

determined, will struggle to prevail unless they are blessed with loyal champions in their clients or employers. Sadly, few of them are. Would Rams's products for Braun have been as exceptional without the support of its empathetic owners, Artur and Erwin Braun? Probably not. Will Apple's be without Steve Jobs? We will see. Most successful examples of commercial design are due not just to talented designers, but to enlightened managers. 'I cannot count the number of clients who have marched in and said: "Give me the next iPod,"' noted Tim Brown, who runs IDEO. 'But it is probably pretty close to the number of designers I've heard respond (under their breath): "Give me the next Steve Jobs."'[10]

One of the best models of corporate design championship was London Transport during the early twentieth century, when the design programme was run by Frank Pick. A lawyer by training, he joined the London Underground in 1906 as assistant to the deputy chairman, and became managing director of the London Passenger Transport Board, when it was founded in 1933. Pick insisted that every aspect of the network should be of the highest quality, including its design. Artists such as Man Ray, Graham Sutherland, Edward McKnight-Kauffer and Paul Nash were commissioned to design posters, as was László Moholy-Nagy during his brief stay in London en route to Chicago.[11] Moholy also produced a booklet explaining how wary Londoners should use the newfangled escalators then being installed in Underground stations. It was under Pick's watch that a young draughtsman, Harry Beck, devised a diagrammatic map to guide passengers around the labyrinthine network as it expanded into newly built suburbs. And it was Pick who charged the typography designer Edward Johnston with the development of the red, white and blue roundel symbol that the London Underground has used ever since.[12]

Critically, Pick realized that if his inspired design commissions were to be effective, their execution and maintenance had to be meticulous. To this end, he devoted his evenings and weekends to travelling the length and breadth of the network, checking that everything was up to scratch. He made a note of every glitch, however minor, and fired off memos the following day, instructing the station and depot managers to remove peeling posters or to mend fraying seat upholstery. Sadly, Frank Pick was exceptional, as in their eras were Josiah Wedgwood, the Braun brothers and Steve Jobs. Few senior executives come close to matching their understanding of design, and often end

The Studio H experimental design course at Bertie Early College High School in Windsor, North Carolina

Research
Ideate
Develop
Prototype

Build
Pencils 72
72 Colored Pencils

Farmer's
New

up adding to the pressures on the designers in their employ by creating yet more obstacles to be overcome.

A cautionary tale of how a well-intentioned design project was torpedoed by organizational inefficiency is that of another public transport system, the New York subway, during the 1960s and 1970s. Up until then, the subway had sported a motley assortment of signs in different styles and sizes, reflecting the fractured nature of the network, which was cobbled together in a series of mergers between different lines. The end result was confusing, as the title of a 1957 design proposal suggested: 'Out of the Labyrinth: A plea and a plan for improved passenger information in the New York subways'. Even so, it took until 1966 for the transit authority to act, by appointing the Unimark design group to review the signage system. Paul Shaw's book *Helvetica and the New York City Subway System* recounts how the Unimark designers watched in horror as the Bergen Street Sign Shop, where the subway signs were made, disregarded some of their proposals and misinterpreted others. The transit authority, which was strapped for cash after a recent strike, refused to allow them to oversee the process, arguing that it could not afford to do so. Eventually it relented, but insisted that the signs be made at Bergen Street and that the typeface had to be one which the workshop already owned. This ruled out Unimark's chosen font, Helvetica, and the similar but clumsier Standard Medium was used instead. Nor could the transit authority afford to buy enough new signs to replace all of the existing ones. The result was another muddle, albeit a milder one. The same problems recurred when Unimark introduced a new design scheme in the late 1970s, and it was not until 1989 that Helvetica, the crisp, modern typeface that suits New York so well, was finally introduced to the subway.[13]

Even worse is when design quality is imperilled by mismanagement across an industry, rather than within one organization. Take fashion, where chief designers and creative directors can be hired and fired with alarming speed by companies that tend to be run by executives whose management skills were honed in other sectors, such as banking or food. Some of them become genuinely knowledgeable about the fashion system, and sensitive to the idiosyncrasies of individual designers, but all too often they try to impose managerial methods and financial disciplines which may work perfectly well in other businesses, but are ill-suited to the fashion world.

Inexperience, shortsightedness and obduracy are

problematic in any business, but can be particularly damaging in fashion, where the chief designer of each label is obliged to produce anything up to a couple of dozen collections each year. They are also locked into a punishing cycle of reinventing their vision of the brand in each collection, and subjecting the results to public scrutiny. Such pressures may have contributed to Alexander McQueen's suicide in 2010, and to John Galliano's breakdown, which culminated in his dismissal from Christian Dior the following year, after being arrested for an anti-Semitic rant in a Paris bar.[14]

Not that Galliano's behaviour would be acceptable under any circumstances, but his fragility and McQueen's tragic death are stark reminders of how vulnerable designers can be, not least as the appeal of many fashion brands is steeped in the romantic mythology of volatile designers and their extreme behaviour. Some designers are robust enough to cope and even flourish. Karl Lagerfeld is known as 'Kaiser Karl' in the industry, partly for the vigour with which he has marketed his skills to different companies over the years and rebuffed would-be rivals; while Helmut Lang was responsible for both business and design at his company before selling it to Prada in 2004 and retiring the next year. But most designers need the support of efficient, empathetic managers, who are hard to find. Tellingly, many of the most successful creative and commercial duos in fashion have been both personal and professional partnerships. Yves Saint Laurent founded his fashion house in 1962 with his then-lover, the former art dealer Pierre Bergé, and Miuccia Prada's business partner is her husband, Patrizio Bertelli. But other fashion brands have suffered from being run in cookie-cutter fashion by corporate executives, who may well be capable managers, but are neither imaginative, nor adroit enough to understand the vagaries of the fashion cycle, or the design temperament. The serial sackings are not only expensive and embarrassing for the companies concerned, but can be bruising for the individuals, all of which detracts from the design quality of the clothes.

Even the best-managed design projects in fashion and elsewhere can be scuppered by external factors, especially unexpected events such as corporate takeovers, economic turbulence, fluctuating commodity prices and exchange rates, or political changes.[15] Good design can also be bedevilled by imbalances of power within particular industries. One reason why using a mobile phone can seem confusing is because of the influence of the cellular networks, which exert tremendous pressure on phone makers to ensure that their operating systems are compatible with the networks' technology. Unless the manufacturers oblige, the networks may refuse to sell their products. As a result, operating software is often designed to appease the networks, rather than to make life easier for the phones' owners. Steve Jobs is said to have delayed Apple's entry into the market with the iPhone for several years because of his concerns about the networks' power, and the risk of callers blaming Apple for glitches in their service, which the company would be powerless to correct. Apple refused to comply with the networks' demands and produced an exemplar of intuitive user interface design for the iPhone. Sadly, few other companies are courageous enough to do the same, and can unwittingly create impediments of their own.

A common pitfall is 'change for change's sake' syndrome. Ambitious but insecure executives are particularly prone to this. Anxious to make their mark in new roles, they conclude that the simplest way of doing so is to effect dramatic change, regardless of whether it is necessary or desirable. This can explain why companies take otherwise incomprehensible decisions to ditch great design projects and to replace them with unworthy successors. Why else would the courier company UPS have dumped the wonderful 'present' logo designed for it by Paul Rand in 1961 – a parcel tied with a bow above the company's historic shield symbol – for a bland, digitally airbrushed shield, memorably described by one design blogger as 'the golden combover'?[16] And why else would Citroën have replaced the upturned Vs, modelled on the herringbone gears invented by its founder, the great engineer André Citroën, with a couple of characterless digital smudges?[17] Presumably the management of those companies believed that their new symbols were either better designed or more eloquent representations of their corporate values than the old ones, but, if so, they were mistaken. As was the fashion designer Hedi Slimane when one of the first changes he made as creative director of Yves Saint Laurent was to rebrand its ready-to-wear collections as 'Saint Laurent Paris' and to replace the beautiful initialled symbol drawn for the house by the illustrator Cassandre in the early 1960s. If Slimane wished to demonstrate that he was imposing his vision on the fashion collections, he succeeded, but at the risk of appearing impetuous.[18]

Another version of the same condition could be called 'designeritis', which often afflicts products in markets that experience sudden sales growth. An example is the espresso machine. Once they were purist, utilitarian, no-nonsense exercises in engineering. But no sooner had Starbucks' success convinced kitchen equipment manufacturers that we might like to make our own iced skinny flavoured lattes at home than they unleashed a monstrous new breed of espresso machines. They are the SUVs of the kitchen. Too big. Too bombastic. Too wasteful. Too much.

Equally vulnerable are elderly products which fulfil their functions reasonably well but have exhausted the possibility of improvement. An example is the toaster. For centuries, people toasted bread on open fires by spearing it on sticks, forks, knives or swords. By the late 1800s, erstwhile inventors were constructing electrical contraptions to toast it on heated strips of iron wire. The hitch was that the wire was prone to melting and bursting into flames. The golden age of toaster design was in the early 1900s, when the machines were made safer and more reliable. The automatic pop-up toaster was patented in 1919 by Charles Strite, a master mechanic working in Minneapolis, and went on sale seven years later as the Model 1-A-1, the first toaster with a timer that could be set to toast the bread on both sides at once, before ejecting it.[19]

Toasters very much like the Model 1-A-1 have been used ever since. There have been some functional improvements, mostly to make them safer and more energy-efficient, if only slightly less likely to spray crumbs. As a result any design innovations have tended to focus on styling, which is why, judging from the toasters on sale in one of the department stores on Oxford Street in London, their designers and manufacturers have succumbed to designeritis.

Why does the DeLonghi Icona toaster appear to require more buttons, dials and levers than the most complex components of the International Space Station? What possessed Bosch to attach what looks like a miniature helipad to the top of its Private Collection Toaster? Or Magimix to think that we have enough time on our hands to watch bread toasting and, presumably, crumbs accumulating through the glass casing of the Vision Toaster? Equally outlandish is the Dualit NewGen, which, despite its name, is so antiquated in style that it looks as if it has been salvaged from the debris of a 1930s Lyons Corner House. While anyone fancying a whiff of futurism could plump for the Bodum Bistro, whose inscrutable case looks likelier to belong inside the Large Hadron Collider than in a kitchen. But if you were hoping to find something reasonably attractive and unobtrusive to toast a couple of slices of bread – tough.

Another product that is prone to over-styling is the chair. We have always needed things to sit on, from conveniently shaped boulders in prehistoric caves and medieval milking stools, to the thrones of monarchs, popes and tyrants. And chairs have acted as excellent canvases for designers over the years, being big enough to use sophisticated materials and production processes, without being too large for designers and manufacturers to baulk at experimentation. They can also be manufactured in huge quantities, thereby justifying the use of complex technologies, or made by hand from precious materials using intricate craft techniques as limited editions or one-offs. 'Every truly original idea – every innovation in design, every new application of materials, every technical invention for furniture – seems to find its most important expression in a chair,' wrote the American furniture designer George Nelson in 1953.[20] Sadly, as more and more chairs have been produced, the possibility of them being either original or important has become progressively slimmer. Every year hundreds of new chairs are unveiled at the Milan Furniture Fair, only a handful of which are distinctive or innovative. Most of the others are forced to resort to stylistic excesses in desperate attempts to attract attention.

Design can also suffer if the context in which it is used changes. Sometimes designers are at fault for failing to anticipate the impact of a change of circumstances on their work or to take effective action to correct it. Website designers are guilty of both offences when they take great pride in showing their clients how stunning digital typefaces like Georgia will look by unveiling them on state-of-the-art Apple computers. What they fail to point out is that the same font will look very different and considerably less elegant on PCs, which are, after all, used by many more people than Apple's machines.

Sometimes the designers are blameless. The Adidas design team must have known that its efforts to produce the best possible official ball for the 2010 World Cup in South Africa would be hampered by FIFA's decision to stage the tournament in cities of dramatically different altitude and air density.[21] After all, similar discrepancies had affected the performance of a different World Cup ball at the 1978 tournament in Argentina, the last one in which the changes in altitude were as stark. The designers of the Jabulani were powerless to prevent it from happening again, because science has not yet succeeded in regulating the performance of a ball moving at speed through such dramatically different altitudes and air densities. The fault lay with FIFA's officials, who must also have been aware of the altitude problem and could surely have revised the fixture schedule accordingly, to minimize the changes in altitude.

Yet all too often design projects flounder because designers have messed up. Sometimes they make misjudgements, like the well-intentioned but misguided designers of a device intended to help people carry water across long distances in Africa. It was a portable water purification unit in the form of a drum that used reverse osmosis, whereby dirty water is cleaned by being strained at high pressure through a membrane. So far so good, except that many of the people who used it had to expend time and energy on rolling gallons of dirty water for miles to extract a few gallons of clean water.

Other designers simply seem not to care. Clearly they would not have been willing to camp out in a 747 for several weeks to complete work on a prototype during the SARS scare. Take my electric toothbrush, a Braun Oral-B Sonic Complete. I bought it because a friend recommended it. After using it for a few months, she was told by her dentist that her teeth were so clean that they did not need to be scaled or polished: a ringing endorsement. My own Sonic Complete turned out to be equally reliable when it came to teeth cleaning, but there its design merits end. Aesthetically, it looks lazy. The shape is clumsy, the mix of colours discordant, the typography ungainly and the silver detailing tacky. It also feels odd to touch, thanks to its incongruous mix of materials.

Even so, I would be willing to overlook those flaws, on the Post-it/Google logo principle of good design, because of the toothbrush's functional strengths, but it is impossible to ignore its environmental failings. The first problem is the box that it arrived in, which was considerably bigger than its contents. The bigger the box, the fewer of them will fit into each freight container, and the more planes, ships and trucks will be needed to transport them, thereby consuming more petrol. Problem two was that the empty space in the box was padded with a type of polystyrene which is neither biodegradable nor easily recyclable. As for what will happen to the Sonic Complete at the end of its working life? Cue problem three. Had Braun made proper provision to dispose of it responsibly? No. The instruction booklet contains a vague suggestion that it should be taken to 'the appropriate collection points provided in your country'. Finally, problem four.

Cleaning my teeth with the Sonic Complete for two minutes twice a day for a fortnight requires sixteen hours of recharging, which adds up to seventeen minutes of recharging for every minute it is used. Hardly energy-efficient.

However pearly white it promises to make my teeth, those niggling shortcomings stop the Sonic Complete from qualifying as good design. Did its designers wake up one morning and decide to design an electric toothbrush which would look ugly, feel awkward, take a preternaturally long time to charge and end up hogging sorely needed landfill space with potentially toxic packaging that has already squandered fossil fuel? Probably not, but that is what they have done.

6 Why everyone wants to 'do an Apple'

It looks like it has come from another planet. A good planet. A planet with better designers.
— Steve Jobs introducing the first iMac in 1998[1]

Before he left, the outgoing chief executive sent a farewell email to all of the staff in which he identified five of the biggest problems facing the company: '1. A shortage of cash and liquidity. 2. Poor quality products. 3. No viable operating system strategy. 4. A corporate culture lacking in accountability and discipline. 5. Fragmentation, trying to do too much and in too many directions.'

The email was written in July 1997 by Gil Amelio, who had spent three gruelling years struggling with the financial problems of Apple Computers only to be ousted in a boardroom coup during the 4 July national holiday. His successor was Steve Jobs, the man who had co-founded the company in 1976 only to be fired ten years later. In his first meeting with his new colleagues, Jobs asked them to 'Tell me what's wrong with this place.' The question turned out to be rhetorical, because he answered it before anyone else could speak. 'The products suck!'[2]

By the time Jobs announced that he was resigning as Apple's chief executive on 24 August 2011, following a prolonged battle against cancer, the business and its products were unrecognizable. Apple's share price had risen so high that at one point it was the world's most valuable company in terms of stock-market value. The news of Jobs's departure was splashed across the front pages of newspapers all over the world. The Internet was flooded by tweets, blog posts and Facebook updates, some saddened by the news, others lauding Jobs's achievements. 'Rarely has a major company and industry been so dominated by a single individual, and so successful,' proclaimed *The New York Times*. 'His influence has gone far beyond the iconic personal computers that were Apple's principal products for its first twenty years. In the last decade Apple has redefined the music business through the iPod, the cellphone business through the iPhone and the entertainment and media world through the iPad. Again and again, Mr Jobs has gambled that he knew what the customer would want, and again and again he has been right.'[3]

A more intense outpouring of grief greeted the news of Jobs's death on 5 October 2011. Mourners arrived at Apple stores all over the world bringing flowers, candles and apples (many with a chunk bitten out, as it is in the company's logo) to leave there in impromptu 'iShrines' to Jobs. The store windows were filled with Post-it Notes on which people lamented his death and celebrated his triumphs.[4] There were even 'iShrines' outside the fake Apple stores that had sprung up in China, laid out exactly like the official ones with staff sporting replicas of genuine Apple T-shirts and name tags.[5] In its obituary, the *Economist* saluted 'his unusual combination of technical smarts, strategic vision, flair for design and sheer force of character' and dubbed Jobs 'the world's most revered chief executive'.[6]

He had died a few weeks after the anti-capitalist activists in the Occupy movement had begun their occupation of Zuccotti Park near Wall Street as the first step in what would rapidly become a global protest against (among other things) greed, corruption and incompetence among the corporate and financial elite. Yet here was one of America's most powerful executives, one whose remuneration package had at times been condemned as excessive, being spoken of not simply with admiration by his fellow executives, but with heartfelt gratitude by the people who had bought his company's products.[7]

Under Jobs's watch, Apple had become the alpha brand of not just one, but several industries. It had done so by designing products which were so desirable that people were prepared to pay more for them than for comparable ones made by its rivals. Company after company talks about wanting to 'do an Apple'. Yet Apple was far from being the first business to reach stellar status by designing products that consumers longed to buy and its competitors strove to copy.

Throughout history, certain things have been deemed so special that they have imbued the people who own or use them with immense prestige. In ancient Rome, only those of or above the rank of senator were allowed to wear clothing dyed in a particular shade of purple. The sole person permitted to wear yellow in Imperial China was the emperor himself. During the Middle Ages the right to wear specific garments was enshrined in legislation, known as the Sumptuary Laws, which became progressively more complex over the years. By the sixteenth century the English aristocracy was so perturbed by the growing wealth of the merchant class that King Henry VIII was persuaded to strengthen the Sumptuary Laws. Ermine, sable and miniver could be worn only by nobles, as, puzzlingly, could broad-toed shoes. No one beneath the rank of knight could wear a silk shirt. Purple was reserved for the king, so was cloth of gold, although dukes and marquesses were allowed to don that too. Nouveaux riches Tudor merchants delighted in flouting the laws, even though they could be stopped on the street if suspected of wearing something taboo and the offending item confiscated as a punishment.[8]

Enforcing such sanctions became increasingly difficult over the years as wealth was dispersed more widely, and impossible after the Industrial Revolution, when the consumer class expanded and the public had more goods to choose from, including ones which were mass-manufactured rather than laboriously made by hand. Until then, the most desirable objects had generally been unique and combined unusually fine or rare materials with intricate craftsmanship. All you needed to acquire them was money and, sometimes, power. Industrialization made consumption seem more democratic by introducing a new genre of alpha products which were not only appealing and of exceptionally high quality, but made in such large quantities that they were affordable, if not to everyone, at least to many more people than the very rich. These products said something different about their owners, and society as a whole, because the decision to buy them was not determined solely by their cost or quality, but also by the purchaser's personal taste.

The first industrial alpha product was Josiah Wedgwood's Queen's Ware, the simply shaped, richly glazed cream-coloured earthenware dinner service that he introduced in 1763. A technical triumph of its day, Queen's Ware was the result of Wedgwood's scientific experiments, as well as the managerial innovations in his factories and his passion for Robert Adam's neoclassical architecture. Those plates, bowls, cups, saucers and gravy boats were among the first mass-manufactured products to be used by teachers and shopkeepers, as well as by royalty, from Queen Charlotte of England to Russia's Catherine the Great.[9] For Wedgwood's newly prosperous middle-class customers, purchasing his products was a seductive opportunity to buy into the values of the aristocracy. They might not have been able to commission Robert Adam to design a mansion, as Lady Templetown and her husband had done,[10] but they could afford to buy a gravy boat in the Adam style, just as people treat themselves to Prada perfume or Céline sunglasses today, when what they really want is an expensive bag or dress that they have spotted in *Vogue*.

Far more than vessels to drink or eat from, Wedgwood's pots offered the tantalizing possibility that industrialization might create a finer, more enlightened society, rather than destroying everything decent about the old one. The German Romantic poet Novalis likened Wedgwood's everyday ceramics to Goethe's poetry: 'He (Goethe) has done for German literature what Wedgwood has for the English art world. Like the Englishman, he has a naturally economical and noble taste acquired through understanding.'[11] Well over a century later, the British art critic Herbert Read analysed their appeal in *English Pottery*, a 1924 book he co-wrote with the ceramics scholar Bernard Rackham: 'Wedgwood was the first potter to think out forms, which should be thoroughly well suited to their purpose, and at the same time capable of duplication with precision in unlimited quantities . . . The shapes are as a rule thoroughly practical and many . . . have remained standard shapes to the present day. Lids fit well, spouts do their work without spilling, bases give safety from overturning, everywhere is efficiency and economy of means.'[12]

Many of the same qualities applied to the alpha products of the nineteenth century: the bentwood chairs introduced in the 1850s by the German furniture maker Michael Thonet and manufactured in the factory he ran with his five sons at Koritschan in what is now the Czech Republic. By then, Thonet had been working as a cabinetmaker for nearly forty years and had struggled to find enough money to keep his business afloat and patent his innovations. Like Josiah Wedgwood, he was a self-made man from humble origins who had prospered by mastering his trade, then reinventing it.[13]

Through his technical experiments, Thonet had found a way of producing smoothly curved yet strong bentwood forms, and identified a simple style of furniture which could be made relatively cheaply and quickly in large quantities. Moving to Koritschan gave him access to cheap labour and wood from the nearby forests as well as good transport links throughout Europe. He and his sons made the most of those advantages by designing furniture to be manufactured from as few components as possible, and by organizing the factory as a model of standardized production with one group of workers dedicated to bending, and others to sawing, steaming, sanding, polishing or packing.

The apogee of Thonet design was the Model No. 14 dining chair, sometimes known as the Consumer Chair, introduced in 1860. Based on Thonet's Model No. 8, which was made by hand from a single piece of wood, Model No. 14 was designed for mass production, and Thonet refined it continuously. By 1861, the chair was manufactured in solid bentwood, which meant that the components could be screwed rather than glued together, and therefore assembled by the user, as flat-pack furniture is today. At a time when most chairs left the factory – or craftsman's workshop – intact, Thonet had a considerable advantage in being able to pack the components of several Model No. 14s into compact wooden crates, which were shipped to their destination, where their contents would be put together. The next step was to reduce the number of components needed to make each chair. By 1867, it was possible to make a Model No. 14 from six pieces of bentwood, ten screws and two washers.[14] As a result the chair cost just three florins, roughly the same price as a cheap bottle of wine, meaning that, like Queen's Ware, it was affordable for teachers as well as aristocrats.

None of this would have mattered if Model No. 14 had not been so appealing. Stylistically, it was perfectly pitched for an era when, nearly a century after the Industrial Revolution, with the Arts and Crafts Movement gathering steam, people were aware of industry's power and eager to take advantage of it, but wary of anything that looked overtly industrial. Model No. 14 was unquestionably a factory product: its curves were too precise to have been formed without mechanical assistance, but they looked more rustic than urban in spirit, as did the gleaming wood in the frame and the woven cane on the seat. In other words, Model No. 14 promised the best of both worlds to its owners. There were enough visual cues to remind them of the handcrafted wooden furniture they still thought of fondly, yet the hint that a machine had been applied to those sleek bentwood curves suggested that the chair should be stronger and more resilient than traditional handmade furniture.

Equally important, Model No. 14 was a pleasure to use, as it is today. So light and compact that it is easy to pick up and to move from place to place, it even improves with age. The patina of the bentwood darkens and deepens over the years, and the screws and glue loosen. When that happens, the frame of the chair slackens just enough so that it seems to relax when you lean back on it. It is subtle changes like this, in colour and structure, that make each Model No. 14 seem very personal to its owner, as if it had been designed and made just for them, rather than as part of a huge production run.[15]

All of these qualities made Model No. 14 immensely popular. Demand was so high that Thonet was soon unable to satisfy it at Koritschan. A new factory was built thirty miles away in Bystritz, and then a third. By 1865, Thonet was manufacturing one hundred and fifty thousand pieces of furniture a year, most of which were Model No. 14s. Those chairs were found in cafés, theatres, hospitals, churches, schools and prisons, and were part of cultural life in Vienna. Johannes Brahms sat on one to compose his music, as did the musicians in Johann Strauss's orchestra while they were playing at imperial balls.

Thonet continued to expand by developing new designs and building more factories, not only to satisfy demand for the furniture but to ensure that the company controlled every aspect of production, even making its own screws and the bricks for its buildings. By the late 1890s, Thonet also provided housing for many of its employees as well as schools, libraries and crèches for them and their families, and shops where they could make purchases using the corporate currency in which their wages were paid.[16]

The Model No. 14's appeal proved to be surprisingly enduring, possibly because being neither wholly industrial nor wholly folkloric, but a little of both, made it too elusive to date easily. Lenin perched on one while writing at his desk in the early 1900s. When Le Corbusier and Pierre Jeanneret had to choose a chair to furnish the model modern interior they had designed for the 1925 Exposition Internationale des Arts Décoratifs in Paris, they plumped for a Model No. 14 too. 'We believe that this chair, millions of which are in use in our continent and the two Americas, is a noble thing,' wrote Le Corbusier, 'for its simplicity is a distillation of the forms that harmonize with the human body.'[17]

Just as Thonet's bentwood furniture embodied the 'efficiency and economy of means' that Herbert Read and Bernard Rackham had so admired in Wedgwood's everyday ceramics, so the alpha products of the mid twentieth century echoed the 'simplicity' and harmony which Le Corbusier had relished in Model No. 14.

They were the radios, record players, shavers, food mixers, juicers and other electronic products made by the German company Braun between the 1950s and early 1980s.

Founded by Max Braun as an electronic components manufacturer in 1921, Braun was taken over by his sons, Erwin and Artur, after his death in 1951. Artur was an engineer, who relished the opportunity to experiment with new technologies, materials and production techniques. Erwin, who had originally intended to become a doctor, was to handle business affairs, and he set about educating himself in that field. In 1954, he attended a lecture given by the industrial designer Wilhelm Wagenfeld, once one of László Moholy-Nagy's students at the Bauhaus.[18]

Dismissing the stylized approach favoured by 'signature' designers such as Raymond Loewy in the United States, Wagenfeld called for companies to integrate design into every aspect of their activities. He also urged them to become 'intelligent manufacturers who thoroughly reflect every product's purpose, its utility and durability' by pursuing a rigorous process of research, checks and cross-checks 'just like a research task in a physician's or chemist's laboratory'. The result, he argued, would be a new genre of products, which were useful, efficient and easy to operate as well as pleasing to look at and use, rather than solely designed to sell. Stylistically, they should appear subtle and unobtrusive, although he ended his lecture with a warning that the simpler a product looked, the more challenging it usually was to make.[19]

Erwin was so impressed that he asked Wagenfeld to advise Braun on design. By the end of the year, the brothers had forged a relationship with a new design school, the Ulm Hochschule für Gestaltung, where Wagenfeld was teaching alongside a gifted young designer, Hans Gugelot. They also began to build their own corporate design team. Early in 1955, Braun advertised for an in-house architect in a local newspaper. A couple of juniors in a Frankfurt architecture office spotted the ad and dared each other to apply. One of them was offered the job: the twenty-three-year-old Dieter Rams.[20]

Rams's first assignments at Braun were architectural projects, but he soon became involved with product design too. Working alongside Erwin Braun, Artur's engineers and the Ulm tutors, he developed and implemented a design system in accordance with the principles outlined in Wagenfeld's lecture. The new regime began with audio products, which were of special interest to the Brauns, who were both jazz buffs. Erwin in particular had a deep love for the music, having belonged to the *Swingjugend* or 'Swing Kids' movement of 1930s German secondary-school students who signalled their opposition to the Nazi regime through their love of jazz, swing dance and dandyism. They sported long hair and adopted the slogan *Swing Heil* to mock the Nazis' *Sieg Heil*.[21]

The brothers were also swayed by practical considerations. They knew that Braun had the technical means to make progress in the audio field by applying the advances in military communications technology made during the Second World War, as well as those of more recent breakthroughs, notably the transistor, which promised to transform the design of electronic products by replacing bulky, unreliable vacuum tubes with something much smaller and more powerful. There was also a promising new market to be cultivated. Musical taste was changing with the growing popularity of jazz, rock 'n' roll and eventually pop. These new genres demanded a different type of sound: cleaner and purer than that of classical music, which existing radios and gramophones had been designed to play. A new style of equipment was needed too. Up until then, audio products seemed to have been designed to disguise the machinery that operated them, by covering it up with carved wooden cases resembling antiques. The outcome looked intentionally old-fashioned.

Why were mid twentieth-century consumers so determined to pretend that their newfangled electrical products were faux antiques? Because they were as intimidated by electronics as Thonet's customers had been by industry a century before? Probably. Braun's triumph was to define a new design language for such products, which was unquestionably modern, yet so well attuned to its users' needs that it felt inviting, rather than cold, harsh or frightening.

The new language is visible in the first product that Rams worked on from start to finish and on which he collaborated with Gugelot and the Ulm team: a 1956 record player and radio known as the SK4 and nicknamed 'Snow White's Coffin'. Modelled on an earlier product designed for Braun by Wagenfeld, the SK4 looked proudly un-antique in its white lacquered sheet-steel case with elm-wood side panels. The buttons and dials were all the same shape and size, and arranged in straight lines, as were the speaker and ventilation slits. To soften the effect, the top of each button was slightly indented, showing the user exactly where to place their finger to push it down. The finishing touch was the turntable lid, which was made in transparent acrylic to ensure that the pick-up arm and audio controls were clearly visible, rather than hidden from view.[22] Years later, Rams recalled how happy he had been to discover that Erwin Braun had credited him and Hans Gugelot with the design.[23] Few other companies of the era would have acknowledged the work of an in-house designer, especially one so young. It was a generous gesture, which showed how important design and its designers were to Braun.

The SK4's icy aesthetic defined the styling of audio products for decades to come, but no company executed it as adroitly as Braun in the hundreds of products developed by Rams and his team. Subtly styled and beautifully proportioned, they were made from technically advanced materials which were as luscious to touch as to look at. Nothing about those machines could be described as showy, yet they had a quietly imposing air of irreproachability, as if every element had been rigorously interrogated, with nothing left to chance.

The same design cues appeared repeatedly. The components of Braun's audio systems were made in identical sizes so they could be stacked neatly on top of each other to conserve space, and to allow customers to upgrade them over time by adding new ones when they could afford them. Edges and corners were gently rounded, as were the button tops, to ensure that touching and operating a Braun device always felt comfortable. Most of the tops were concave, as they were on the SK4. But when electronic buttons were introduced to Braun's pocket calculators in the 1980s, the designers made their tops convex, rather than concave, because it was more important for the users to press them at the correct point than to push them down firmly.[24]

Materials were always immaculate. Artur Braun and his engineers constantly developed new formulas and finishes to make the company's products more durable, and more attractive to look at and touch. Braun was way ahead of its rivals in appreciating the importance of the invisible, intuitive aspects of a product to the people who would use it. Often the choice of surfaces and textures had a functional purpose, as well as sensual and aesthetic ones, like the casing of the Micron Vario 5 shaver, which combined hard and soft plastics to make it easier to grip, even when wet.[25]

Colours were discreet. Believing that everyday products

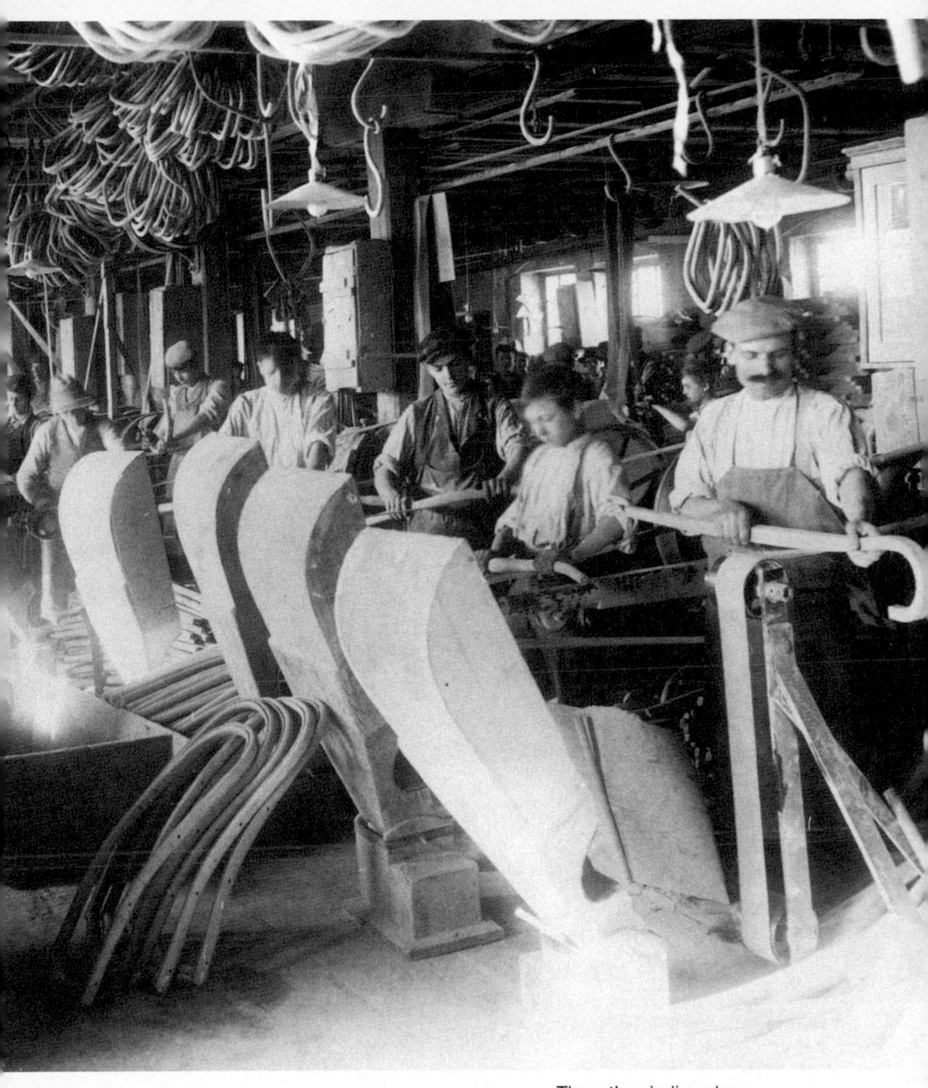

Thonet's grinding shop
in Bystritz, *c.*1920

(next page)

Thonet furniture in a warehouse
in Marseille, *c.*1920

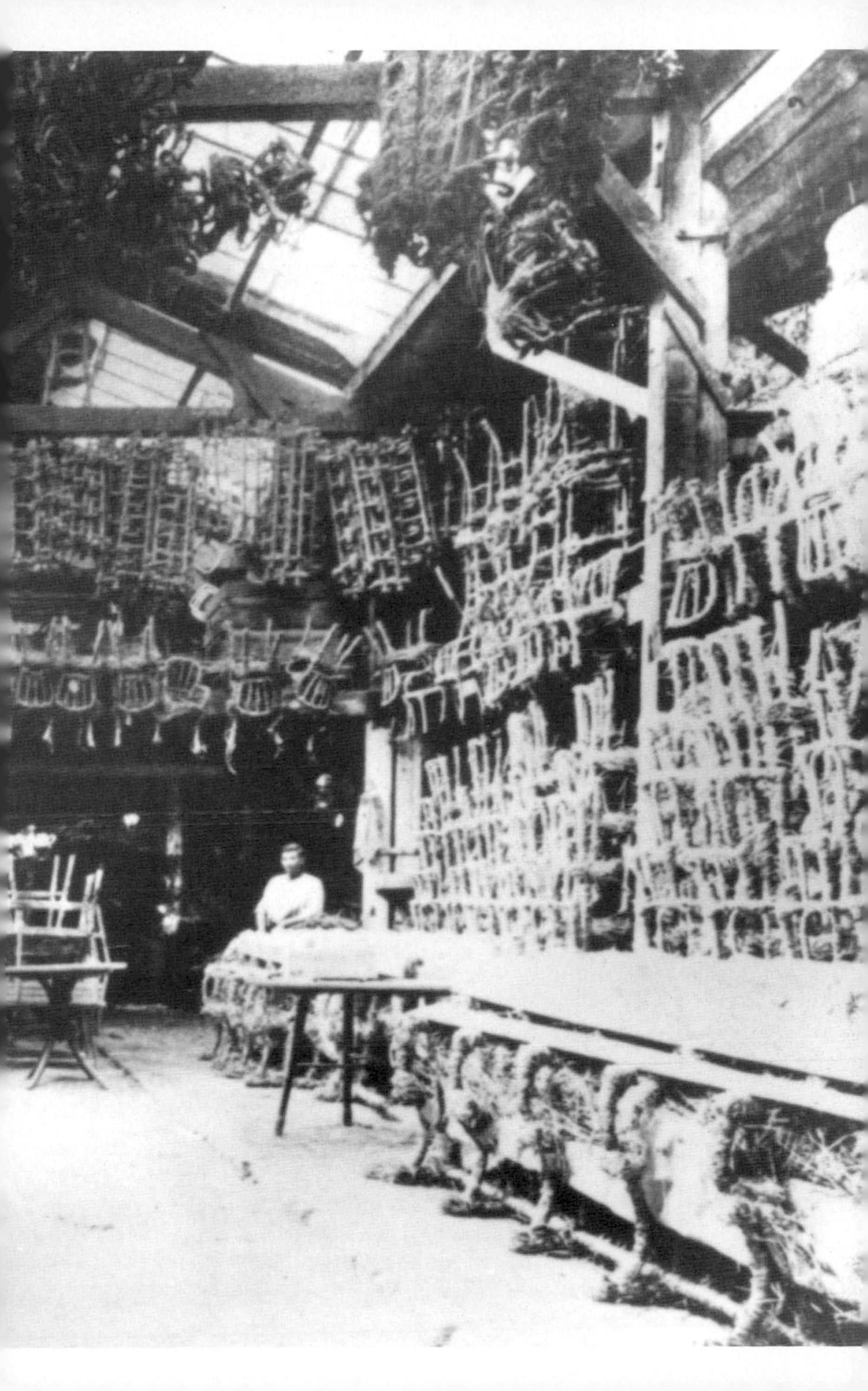

Transporting Thonet furniture
in Bystritz, *c.*1920

like Braun's would be part of people's lives for so long that they 'must stay in the background' and 'harmonize with any environment',[26] Rams favoured whites, pale greys, blacks and metallic hues as base shades. Brighter colours were usually used for guidance, and rigorously coded to help people to operate the products correctly, an important issue for a company that prided itself on introducing new technologies with which its customers would not necessarily be familiar. Red signified the 'off' switch, green 'on', and so on. Even the order of the switches, buttons and dials was determined to appear clear and logical. 'I never trusted instruction manuals,' said Rams. 'We all know that most people don't read them.'[27] The solution was to make the products seem self-explanatory.

A critical component was the design of product graphics, a relatively new field in the 1950s and 1960s. The graphic designers of the era were not trained to work with three-dimensional objects, nor were product designers with graphics. Rams solved the problem by creating a screen-printing facility, run by one of the designers, Dieter Lubs, who had had some experience of graphics earlier in his career as an apprentice in a shipbuilder's design office. Lubs and his team experimented with ways of improving the quality of the product graphics, always using the same typeface, Akzidenz Grotesk, which was a forerunner of Helvetica and one of the clearest fonts then available.[28]

My favourite example of Braun's design purism is the template for its alarm clocks. The second hand was always yellow, and both the alarm hand and the button for setting the alarm were green. The sound of the alarm was identical for every model, as was the position of the hands whenever they were photographed: at eight minutes past ten with the second hand on the six. Not only was this the time at which Dieter Lubs had calculated that the clock faces looked most attractive, the consistent positions of the fingers made the images more memorable, even to people who had glanced at them fleetingly.[29]

And my favourite description of a Braun product was written by Jonathan Ive, Apple's head of industrial design, as the foreword to Sophie Lovell's 2011 book on Dieter Rams. 'It was white, it felt cold and heavy,' he wrote. 'The surfaces were without apology bold, pure, perfectly proportioned, coherent and effortless. There was an honest connection between its blemish-free surfaces and the materials from which they were made. It was clearly made from the best materials, not the cheapest. No part appeared to be hidden or celebrated, just perfectly considered and completely appropriate in the hierarchy of the product and its features. At a glance, you knew exactly what it was and how to use it . . . It felt complete and it felt right.'[30]

Ive was writing about the first Braun product he had encountered – an MPZ 2 Citromatic juicer bought by his parents in the 1970s – but he might have been describing the products that he and his colleagues had developed for Apple: the ones that Steve Jobs once described as having come from 'a planet with better designers'.[31] Jobs had expressed his admiration for Braun as long ago as 1983 by telling the International Design Conference at Aspen that he intended to make Apple's computer's 'beautiful and white, just like Braun does with its electronics'.[32] When he met Ive fourteen years later, they discovered a shared passion for Rams's work there. Eventually, they expressed it publicly by designing the digital keypad of the calculator on the first iPhone in 2007 as a replica of the ET44 calculator developed for Braun in the late 1970s by Rams and Dieter Lubs. The same interface has since been used by every new model of the iPhone.

Not only did Apple and Braun share design objectives, they were run by people with a similar ethos. Both Erwin Braun and Jobs were exceptionally demanding and decisive leaders, steeped in the counter-culture of their times. One in the subversive *Swingjugend* movement of Nazi Germany, the other in the geeky-hippy community of meditation, veganism and yoga of 1970s northern California. Critically, the two companies emerged at watershed moments for consumer culture. Just as Braun used design to convert the public to the magic of electronics, Apple did the same for computing.

Apple was co-founded by two Silicon Valley tech geeks, Jobs and Steve 'Woz' Wozniak, who had been friends since their teens. A brilliant programmer and hacker, Woz was devoted to technological progress;[33] whereas Jobs was an irrepressible entrepreneur with an instinctive understanding of design, having grown up in one of the modern houses built by the Californian developer Joseph Eichler during the 1950s with the aim of making decent architecture affordable to ordinary Americans.

Their first computer, the 1976 Apple I, looked like a hobbyist's machine in its makeshift wooden case,[34] and Jobs was determined that its successor would be more polished. For the Apple II, he asked a local designer to produce a case in light moulded plastic, similar to those of the Cuisinart food processors he had admired in the local branch of Macy's department store.

Jobs agreed to pay $1,500 as a design fee, which he and Woz could ill afford. A colleague told Jobs's biographer Walter Isaacson that as well as spending days agonizing over the precise shape of the machine's corners, Jobs had to be dissuaded from commissioning a special shade of beige for the plastic, and settle for one of the thousands of beiges which were already available.[35] This was a man who lived for years in a house with very little furniture, because he could not find anything that he liked enough to buy. 'The case was really a key,' explained Jobs in an interview with *Playboy* in 1985. 'The real jump with the Apple II was that it was a finished product. It was the first computer you could buy that wasn't a kit.'[36]

Once Apple was up and running, Jobs read up on design history, particularly on Japanese minimalism and the modern movement. He hired the German designer Hartmut Esslinger of frog to work on Apple's products in 1982, and championed his development of a 'Snow White' design language, characterized by uncomplicated shapes and pale colours.[37] Jobs summarized his own design objectives in his Aspen talk the following year: 'Let's make it simple. Really simple' – though he also stressed that simplicity was as important to the design of the way Apple's products would be used, as it was to how they looked. 'The main thing in our design is that we have to make things intuitively obvious.'[38]

Jobs's insistence on usability was evident as soon as Apple's 1984 computer, the Macintosh, was switched on and the words 'Hello' and 'Welcome' flashed up on the screen. By contrast, the first thing people saw when they switched on an IBM PC was 'C:\>', a prompt to the computer's MS-DOS operating system, which would have been instantly recognizable to programmers, but not to anyone else.[39] Though Jobs astonished Apple's engineers with some of his demands, notably by insisting that they redesign the printed circuit board holding the memory chips in the Macintosh, claiming that the original was ugly and had to be replaced. Dismissing their protests that no one would see it buried inside the machine, Jobs recounted how his father, Paul, a keen amateur carpenter, had prided himself on ensuring that every part of his cabinets was made to the same high standard of workmanship, including the hidden ones. He also refused to allow a fan to be added to the Macintosh on the grounds that it could be noisy and distracting, despite the engineers' warnings that the machine would be prone to over-heating without one. Unfortunately they were right, and the Macintosh was nicknamed the 'beige toaster'.[40]

After being ousted from Apple in 1986, Jobs's pernicketiness became even more extreme at his new company, NeXT. Walter Isaacson recounted how he commissioned Paul Rand, the renowned graphic designer who was responsible for IBM's stripy logo, to create a visual identity for NeXT, only to wrangle with him over the shade of yellow chosen for the 'e' in the company's name. Jobs wanted a brighter hue, but Rand, a doughty adversary, refused to budge, banging his fist on the table and barking: 'I've been doing this for fifty years and I know what I'm doing.' Jobs relented, only to clash with Rand again over the position of the '.' in 'Steven P. Jobs', on his business card. Rand had placed the '.' to the right of the 'P', as if it was printed in traditional lead type, whereas Jobs wanted it to be tucked beneath the curve of the 'P', as it would be in the digital typefaces that appear on computer screens.[41]

When he returned to Apple in 1997, Jobs refined his approach to design in collaboration with the British designer Jonathan Ive, who had joined the company a few years before, but would flourish under him, playing Dieter Rams to his Erwin Braun. The first major product they developed together was the brightly coloured 1998 iMac G3. Encasing a computer in vibrant shades of blue, orange, pink or purple plastic seemed outlandish at the time, when most machines were swathed in the drab combination of grey and beige that fashion designers called 'greige' and the novelist Tom Wolfe dubbed 'veal grey'.[42] Riskier still was the use of translucent plastic, through which the innards of the machine were clearly visible. Industry convention suggested that people would be confused at best and frightened at worst by spotting the intricate web of components inside a computer, which is why they were hidden behind opaque plastic, just as the workings of post-war gramophones and radios had been disguised by 'antique' wooden cabinets.

Yet those design decisions made perfect sense given that Apple hoped to sell the iMac to the *Wired* generation of tech-savvy young consumers who had grown up playing video games, often used computers at work and were becoming increasingly reliant on email and the Internet. The bright colours told them that the iMac was cute, fun and fashionable. And the glimpses of its innards should reassure them that using a computer, at least one like this, might not be as tricky or scary as they feared. (Even the niggliest of Apple engineers must have understood why Jobs wanted the inside of the iMac to be as immaculate as the outside.) Equally encouraging was the handle that invited you to pick up the machine. As Ive explained years later: 'Back then, people weren't comfortable with technology. If you're scared of something then you won't touch it. I could see my mum being scared to touch it. So I thought, if there's a handle on it, it makes a relationship possible. It's approachable. It's intuitive. It gives you permission to touch.' Critically, Jobs insisted that Apple signed off on the iMac's colourful casing even though the engineers pointed out that each one would cost over $60 to make, three times more than a standard computer case.[43]

The iMac was a commercial triumph and signalled the start of Apple's revival under the Jobs–Ive design regime. It set a precedent for the user-friendliness of future Apple products, but it took another two years for the two men to define a visual style for the company. The breakthrough product, the 2000 Power Mac G4 Cube, was not a commercial success, although it was a personal favourite of Ive's. An eight-inch box without a single button, it took his and Jobs's obsession with simplification to extremes. Stylistically, the Cube was 'bold, pure, perfectly proportioned, coherent and effortless', all of the qualities that he had admired in his parents' 1970s Braun juicer.[44] The glacial palette of whites and greys, the simple rectangular forms with gently curved edges and corners and subtly contrasting clear and opaque plastics with immaculate finishes made it seem as refined as the iMac G3 was playful. Those stylistic cues would be identified with Apple from then onwards as Ive's design team applied them to new types of products, including the iPod, iPhone and iPad. The visual language was so strongly associated with Apple by the time the iPod appeared in 2001 that it could be recognized from its tiny white earphones if the device itself was hidden inside a bag or a pocket.

Apple's design goals during the Jobs–Ive era were summarized by one of the iPad 2's promotional slogans: 'Thinner. Lighter. Faster.'[45] Whenever a new model of an existing product was developed, whatever else it did, it generally checked those boxes. But the overriding objective was to produce what Jobs jubilantly called 'insanely great' products, ones so sought-after that their launches were presaged by months of rumours and counter-rumours on the blogosphere, prompting thousands of people to queue outside Apple stores to buy them on their first day on sale. How did Apple do this?

One tactic was to ensure that every area of the business – design, product development, engineering, production, finance, distribution, and so on – was focused on developing the best possible products. This sounds simple and something that any sensible company should do, yet depressingly few of them pull it off. Ideally, the different areas of any business should work together productively, but all too often the demands of, say, the engineers conflict with those of the finance team or the designers. Such clashes are common in every company. Most resolve them by prioritizing financial concerns, something that Jobs refused to do. The few companies that are determined enough to avoid this generally end up allowing the engineering team to set the product development agenda, particularly if, like Apple, they are in the business of producing complex technical devices such as computers and phones. Typically the engineers determine what type of products the company will develop, before the designers are drafted in to make them appealing and user-friendly.

Again and again, Jobs reversed this process at Apple by championing design over engineering. He believed that he and the design team were best placed to identify what type of 'insanely great' product Apple should produce next, and that the engineers' role was to come up with the necessary technologies to meet their needs, not the other way around. When the Macintosh was being developed, he signed off on the case and instructed the engineers to ensure that the components fitted inside it. During his second stint at Apple, Jobs's attachment to design intensified as his rapport with Ive strengthened. Walter Isaacson described how he regularly supported Ive in his battles with Jon Rubinstein, an old friend of Jobs's from NeXT who joined him at Apple in 1997 to run the hardware division. Most of the disputes were triggered because Rubinstein felt that Ive was too demanding of the engineers. Jobs promoted Ive to report directly to him rather than to Rubinstein, but the rows continued. When Ive eventually issued a 'him or me' ultimatum, Jobs chose him, and Rubinstein left Apple in 2006.[46]

Jobs was a regular visitor to the Apple design studio, which is shielded from the outside world by tinted windows and a locked door on the ground floor of Two Infinite Loop on the corporate campus at Cupertino. Even senior executives need special permission to enter, except for Jobs, who dropped in several times a day. He often lunched there while discussing the progress of different projects with Ive and the designers, and frequently returned later in the day, thereby ensuring that he was fully informed about the trajectory of each product.[47] Whenever problems arose, Jobs would be told about them immediately, just as he would be aware of niggling nuances to be addressed or of any promising developments. Rather than scrutinizing completed proposals or prototypes as other chief executives tend to do, Jobs could have his say throughout the development process. He was, as Malcolm Gladwell wrote in the *New Yorker* after his death, the 'greatest tweaker of his generation'.[48]

It would have been impossible for Jobs to have tweaked so adroitly if he had not had such an intelligent understanding of design, in stark contrast to other senior executives. 'We don't have a good language to talk about this kind of thing,' he wrote in an article published in *Fortune* magazine in 2000. 'In most people's vocabularies, design means veneer. It's interior decorating. It's the fabric of the curtains and the sofa. But to me, nothing could be further from the meaning of design. Design is the fundamental soul of a man-made creation that ends up expressing itself in successive outer layers of the product or service. The iMac is not just the colour or translucence of the shape of the shell. The essence of the iMac is to be the finest possible consumer computer in which every element plays together.'[49]

Critically, Jobs understood the importance of design in determining the experience of using the product, and ensuring that it would feel pleasurable and empowering, rather than forbidding. 'We don't really talk about design a lot around here, we talk about how things work,' he said a year after the iPod's debut. 'Most people think it is how they look, but it is not really how they look it is how they work.'[50] Jobs knew that unless Apple's customers were confident that they could use a new product like an iPod or iPhone instinctively, they would not buy it, just as they would have passed on its earlier models. 'This is what customers pay us for,' he explained in *Fortune*, 'to sweat all these details so it's easy and pleasant for them to use our computers.'[51] The engineers labouring on the iPod's software soon realized that if more than three clicks were needed to perform a particular function, Jobs would insist that they found a faster alternative. The same thing happened if he felt uncertain at any point in the process. He would not be satisfied until operating the device felt effortless.[52]

Another of Jobs's strengths as a design champion was a refusal to accept anything less than 'insanely great'. Other chief executives made similarly grandiose claims, especially when talking to the media, but few of them delivered. Jobs was unusual in having the courage of his convictions and refusing to compromise, even if it threatened to be expensive. When the template for the first Apple store was about to be finalized he sanctioned a six-month delay because a colleague convinced him that they needed to rethink the layout by displaying the merchandise according to how the customers might use different products, rather than grouping them in conventional categories like other stores.[53]

Equally exceptional was Jobs's willingness to take such decisions without recourse to market research, consumer focus groups or other corporate crutches. He was a fervent proponent of the belief that if you ask people what they want, they will describe something that they already know, which is why so many products are doomed to compromise and blandness. And he had absolute confidence in his own ability to play the constructively critical role of Apple's super-consumer. 'It doesn't mean that we don't listen to customers,' he told *Fortune*, 'but it is hard for them to tell you what they want when they've never seen anything remotely like it.'[54] Nor was he afraid of releasing products which were more expensive than those of Apple's rivals. The first iPods cost $399, so much more than other MP3 players that bloggers claimed the letters stood for 'Idiots Price Our Devices'.[55] Even if they had, the hefty price did not deter people from buying them.

Mostly, Jobs's gutsiness paid off, but occasionally it did not. Sometimes a product misfired, as the Power Mac Cube G4 did, and the MobileMe data sharing system, which he described as 'not our finest hour' in an email to employees.[56] And sometimes he and Ive were wrong to ignore the engineers' advice, such as the time when they insisted on using solid brushed aluminium on the edge of the iPhone 4 despite warnings that it could compromise the antenna. It did.[57]

These setbacks pale beside Apple's triumphs under Jobs, raising the question of whether it can sustain its success without him. Perhaps it will. After all, Apple's design coups owed at least as much to Ive and his team as to Jobs himself, as well as to the logistical prowess of Tim Cook, who succeeded him as chief executive. It is also arguable that Apple would have had to revise its design priorities in any event, not least to meet its customers' rising expectations of ethical and environmental responsibility, which were among its weaker points during the Jobs era.

Spectacularly successful though Apple's design strategy was, it had started to look dated even before Jobs's death. With the exception of its timely emphasis on usability, many of Apple's design assets, particularly the luscious styling of its products and carefully orchestrated stream of 'new' models, were steeped in twentieth-century design values, as was its apparent indifference to the criticism of its environmental record and complaints about the working conditions and safety standards in its subcontractors' factories. Many of its competitors faced similar accusations, but the critics singled out Apple on the grounds that it should be held to higher standards, as a business which was perceived – not least by itself – as an industry leader.

Typically, companies like Apple take investment decisions by balancing their profit expectations against the cost of delivering products with the attributes that are likeliest to appeal to their customers. During the Jobs era, performance, styling and ease of use were the chief concerns, together with price, even though Apple's devices were comparatively expensive. As consumers, we make similar calculations when deciding whether or not to buy, say, an iPad or iPhone, assessing the relative merits of its potential usefulness, appeal, cost, and if anything about it gives us cause for concern. The importance of each of those factors to us changes constantly, and by the time Jobs died in 2011, more and more people were becoming averse to buying products if they suspected that some aspect of their production had damaged the environment, or that the workers who had made them were being exploited. Apple was in danger of being found wanting in an increasingly important area of design, and of alienating the growing number of consumers who were questioning its integrity. Up until then, it had been fortunate in that its rivals had failed to match its design strengths, but suddenly it faced a race to redress its weakness on the ethical and environmental front, lest one of its competitors did so first.[58]

Should Apple falter, whether it is for that reason or any other, it will lose its alpha status as a design brand, just as Wedgwood, Thonet and Braun eventually did. And if that happens, which company will produce the next genre of products or services that consumers crave and its competitors long to copy? Whoever it turns out to be, Apple's successor is likely to share many of the qualities that have characterized past alpha design brands.

One is that it will be run by people who are passionate about its products, and immensely knowledgeable about how they are made and will be used. Josiah Wedgwood and Michael Thonet both learnt the practical side of their trades as teenage apprentices, but had the vision and chutzpah to found their own businesses and redefine those industries. Neither Erwin Braun nor Steve Jobs could match their technical prowess, but they had the enthusiasm and knowledge required to spot people with those skills. Both Dieter Rams and Jonathan Ive discovered the pleasure of making things as children: the former from his grandfather, a master joiner; and the latter from his father, a silversmith, who often took him to the workshop in the school where he taught. Jobs, too, worked alongside his father in the garage of their home, helping him to restore old cars as well as with his carpentry. These experiences taught them to take pride in their companies' products, and to find joy in the experience of making them, all of which was evident in the outcome. 'As designers it is easy to get so far removed from the actual object,' said Ive in the 2009 feature documentary *Objectified*. 'You can design virtually, prototypes can be made remotely, the actual product is manufactured in another country. It used to be that it was manufactured downstairs, you would develop the product in such a fluid, organic way with how it was going to be made, and that doesn't happen any more.'[59] Much has been written about how handcrafted objects can be imbued with the love, pride and sheer pleasure that characterized the experience of making them, but exceptional design can instill the same qualities in industrial products.

Wedgwood, Thonet, Braun and Jobs also exhibited entrepreneurial ingenuity, strategic vision, courage and perfectionism throughout their careers, though those characteristics are essential in any area of business, not just in developing exceptional products. Josiah Wedgwood devoted years to experimenting with ways of producing his most demanding ceramic pieces, and Jonathan Ive cited Apple's willingness 'to throw something away, because it's not great and try again' as one of its biggest strengths.[60] No other company is likely to produce anything as remarkable unless it is brave enough to do the same.

Alpha products must also mean more to the people who use them than things that fulfil their function efficiently, and are appealing additions to their lives. They need to forge an emotional bond with their users by acquiring symbolic importance. For Wedgwood's cups, saucers and plates, it was the promise that mechanization could make life more civilized at the start of the Industrial Revolution, just as it was for Thonet's bentwood chairs a century later when industry seemed ubiquitous, but bleak and destructive. Braun's beautifully resolved electronic devices offered the same possibility for a different generation of technology in the mid twentieth century, as did Apple's digital products, first for computers and then for multimedia devices.

In this respect, all four companies benefited from the good fortune of operating at a time when breakthroughs in science and technology had the potential to transform daily life. But they flourished because it was they, rather than their rivals, that had the intelligence and ingenuity to fulfil design's elemental role as an agent of change by translating those innovations into objects which were useful or desirable, ideally both. Every consumer electronics company stood to benefit from the invention of the transistor in the early 1950s, but none pulled it off as adroitly as Braun. Twenty years later, the engineers at Xerox PARC in Palo Alto invented the technologies needed to develop a legible graphical user interface, yet it took Apple's design prowess to package their research coups in a machine which was compelling enough for people to want to buy it.

The next alpha company will do the same, though one thing is for sure, it will not aspire to 'do an Apple'. Apple triumphed by being resolutely itself, not by copying someone else, as did Wedgwood, Thonet and Braun, and so will their eventual successor.

7 Why design is not – and should never be confused with – art

There should be no such thing as art divorced from life, with beautiful things to look at and hideous things to use. If what we use every day is made with art, and not thrown together by chance or caprice, then we shall have nothing to hide.
— Bruno Munari [1]

Every so often, artworks are accidentally damaged while being shipped from place to place, which is what happened to a piece by Richard Hamilton when it was returned to his London studio in 1968 from an exhibition in Germany. By the time it arrived, it had been smashed to bits. Luckily, the work was insured, but when the German insurance company was informed, it refused to pay up, insisting that the debris – scratched fragments of aluminium and Perspex, tied together with string – could not possibly belong to a work of art.[2]

Perhaps the knowledge that the piece was called *Toaster*, and that its remains were originally used to create a partial replica of a Braun HT 2 single-split toaster, contributed to the insurers' scorn. Hamilton had made it as an homage to the beautifully resolved electrical products developed for Braun by Dieter Rams.[3] He admired Rams greatly, once claiming: 'his consumer products have come to occupy a place in my heart and consciousness that the Mont Sainte-Victoire did in Cézanne's.' Eventually the insurance company relented, and Hamilton remade the piece.[4]

The saga of *Toaster* and its sceptical insurers serves as a cautionary tale for the traditional view of the relationship between art and design, one that Richard Hamilton and Dieter Rams were anxious to dispel, but which persists to this day. Countless texts have been written on the subject, but most of their content can be summed up in six words and two symbols: art = good, design = not so good.

Art, or so the traditional argument goes, is intellectually

superior to design, because artists are free to express whatever they wish in whichever form they choose, often, but not always, in work they make themselves. Whereas designers are encumbered by numerous constraints – from the need to meet their clients' demands and to ensure that their work fulfils a specific function, to their decision to forfeit control of the outcome by delegating production elsewhere – all of which is generally interpreted as serving them right for sullying themselves with grubby commerce.

This is, of course, a simplification, but there is some truth to it. Many designers are inhibited by those very constraints, and the weaker ones fail to overcome them, while gifted artists exercise their right to self-expression by articulating ideas and emotions that the rest of us would struggle to voice. Talentless artists do nothing of the sort, though that is another story, as is the readiness of mediocre artists to pander to the commercial demands of the market.

Flawed and over-simplified though it is, the 'art = good . . .' argument has proved infuriatingly robust, and design is routinely treated like a 'poor relation' by the art establishment. A design curator at a museum of art and design recalled a colleague from the art department suggesting that they should acquire some of Donald Judd's furniture for the design collection. The design specialist replied that as Judd was considerably more accomplished as an artist than as a designer, surely his work belonged in the art collection and should be paid for from its budget. After all, if a gifted designer had made uninspired paintings, would the art department be willing to buy them? Of course not. So why should the work of an artist automatically be deemed to have greater cultural value than a designer's?

Why indeed, especially as the historical rapport between art and design was considerably more complex and fluid than the 'art = good . . .' stereotype suggests, and is even more so now.

Originally there was no distinction between art and design. In ancient Greece, the same word *technĕ* was applied to both disciplines as well as to the mastery of all other crafts, plus medicine and music – though it would be wrong to interpret this as implying that the ancient Greeks held design in as high esteem as we now do art, because the reverse was true. Painting and sculpture were dismissed as lowly occupations, as were all theoretical forms of *technĕ*, including the process we now identify as design. The only reputable applications were considered to be the practical ones, which is why Plato repeatedly praises craftsmen over artists in *The Republic*.[5]

During the Renaissance, artists elevated their status within society but still engaged in other disciplines, as Giorgio Vasari discovered when visiting Leonardo da Vinci's studio in mid sixteenth-century Florence. Vasari noted the 'many architectural drawings', 'models and plans showing how to excavate and tunnel through mountains without difficulty, so as to pass from one level to another' and drawings of contraptions 'to lift and draw great weights by means of levers, hoists and winches'. 'His brain was always busy on such devices,' he wrote, 'and one can find drawings of his ideas and experiments scattered among our craftsmen today.'[6] Leonardo was exceptionally eclectic, even for the time, although his contemporaries' interests also extended beyond painting and sculpture. Andrea del Verrocchio, the artist to whom he was apprenticed, was renowned for his work in architecture, goldsmithery and carpentry.[7]

Yet when the first art and design schools opened in Italy in the late sixteenth century, art and design were often taught separately. The students at the very first one, the Accademia e Compagnia delle Arti del Disegno, which was founded in Florence in 1563, largely thanks to Vasari's championship, studied art in one branch and design in another. The Accademia di San Luca went further after opening in Rome in 1577 by according higher status to the arts than to craftsmanship.[8] When other art academies emerged across Europe in the seventeenth and eighteenth centuries, including the Académie des Beaux-Arts in Paris, they did the same.[9] As well as educating young artists and architects, these schools acted as forums where practitioners could discuss important issues and express collective concerns. By doing so, artists and architects became progressively more influential within society, especially when compared to craftsmen, who continued to train and work in the traditional way, often isolated within their workshops.

Immediately after the Industrial Revolution, the concept of design was still evolving, as was its relationship to art. Prominent artists and art historians shared the general fascination with dynamic enterprises like Josiah Wedgwood's. This was an era when factory machinery was praised for its power and sophistication, with prizes awarded for the most impressive examples, which were exhibited in 'practical displays' where the public could scrutinize them.[10] Industrial design was by no means deemed to be on an equal footing with fine art as it was then called, but was considered at its best to have intellectual depth and its own allure.

Decades later, industry had not only lost its early cachet but found itself being demonized. The Arts and Crafts Movement's revival of pre-industrial craftsmanship had a lasting influence, especially in Britain and the United States. The organizers of the 1851 Great Exhibition in London hoped that the public would be seduced by the industrial marvels displayed inside the Crystal Palace. But when the future Arts and Crafts evangelist William Morris was taken there in his teens by his parents, he refused to enter, convinced that he would hate its contents.[11]

Conservative though it was, the Arts and Crafts Movement did at least set an encouraging precedent of embracing different disciplines, even if it rejected anything with a whiff of industrialization. The late nineteenth-century movements of art nouveau in France and Belgium, Jugendstil in Germany and the Vienna Secession in Austria were equally eclectic and less hostile to industry, if far from enthusiastic about it. But by the early twentieth century, attitudes had changed dramatically. In Eastern Europe, avant-garde artists, designers and architects united under the banner of Constructivism. And in Germany, the formation of the Deutsche Werkbund represented a concerted effort both to raise standards of industrial design and to generate intellectual debate about it.

The influence of Constructivism and the Werkbund paved the way for the foundation of the Bauhaus, which opened in the German city of Weimar in 1919, a year after the end of the First World War. Billing itself as a new type of art and design school, the Bauhaus promised to 'embrace architecture, sculpture and painting in a new unity' in the 'manifesto' written by its founding director, the architect Walter Gropius.[12] It is now seen as a progressive institution which championed design, performance and photography, as well as the disciplines cited in the manifesto, and encouraged its students to work together to build a fairer, more dynamic society. But in reality, it took time for that ethos to emerge. Gropius was a gifted publicist, and thanks to his efforts we now think of his school as having been visionary, egalitarian and technocratic from the start. But the early years of the Bauhaus were characterized by what the textile designer Anni Albers, who arrived there in 1922, was to describe as 'a great muddle'.[13]

Many of the students and teachers, including Gropius

himself, had fought in the German army during the First World War and had yet to recover from the trauma, physically or emotionally. Few of the original staff survived the first two years of the school's existence. Women students protested against being confined to the weaving and ceramics workshops. Gropius faced constant complaints from local residents about the students' rowdiness, as well as accusations by the increasingly powerful Nazi Party that the Bauhaus was a hotbed of Bolshevik subversion. But his biggest problem was the growing influence of a teacher, the Swiss artist Johannes Itten, a charismatic member of the Mazdaznan religious sect. Clad in flowing robes with a shaven head and smelling strongly of garlic, Itten preached the merits of vegetarianism to the students and urged them to begin each class with breathing exercises. He also imbued his charges with his own vision of art, which was instinctive, visceral, laced with spirituality and far closer to the Arts and Crafts Movement than to Constructivism.[14]

A power struggle ensued, and it was only after Itten's departure in 1923 that Gropius was able to reorientate the school, helped by a new recruit, the overalls-clad László Moholy-Nagy, who fired his charges with his passion for technology. Female students were given a wider choice of courses. The Bauhaus adopted a new slogan, 'Art and technology: a new unity', and the teachers were encouraged to prepare their charges to design for industry.[15] Two years later, the school moved to purpose-built premises in Dessau, designed by Gropius in the modernist style. After his departure in 1928, the architect Hannes Meyer took over as director, followed by Mies van der Rohe in 1930, but continued opposition from the Nazi Party forced the school to close in 1933. By then, many of the greatest names in twentieth-century art, architecture and design had taught or studied there: Josef Albers, Wassily Kandinsky and Paul Klee in painting; Anni Albers and Gunta Stölzl in textiles; Herbert Bayer and Moholy-Nagy in communications; Marianne Brandt, Marcel Breuer and Wilhelm Wagenfeld in product and furniture design; Oskar Schlemmer in performance; Mies van der Rohe and Gropius in architecture. Many of them collaborated with fellow Bauhaüslers from different disciplines at the school, and forged firm friendships. Some, like the Alberses, even married.

The inclusive, collaborative vision of visual culture taught at the Bauhaus, at least during the 'Art and technology' era, had a profound influence, not only on the teaching of art and design

The unsung design ingenuity of the customized tricycles in Beijing used as trucks and mobile workshops

他人

Arthur

LUO
洛宝贝
阳
推进科

砂锅 炒菜 米饭 面条

but also on public perceptions of them, in Germany and elsewhere. Not for nothing had Gropius and his colleagues participated in international conferences and welcomed visitors to the school from all over the world. Among them was an American architecture student, Philip Johnson, who visited the Bauhaus in Dessau in 1930 at the suggestion of Alfred Barr, the founding director of the Museum of Modern Art in New York. Johnson wrote excitedly to his mother that the Bauhaus was: 'magnificent . . . the most beautiful building we have ever seen.'[16] After returning to the United States, he agreed to join Barr at the museum, where he was charged with setting up an architecture department. Once in place, he took the bold step of adding design exhibitions to the programme, including the 1934 show 'Machine Art', which celebrated the utilitarian beauty of ball bearings, gears, pistons, springs, wire rope, propellers, furnaces, pots, pans and other industrial artefacts. Barr began the catalogue with a quote from Plato, and Johnson appointed a panel of 'experts', including the aviatrix Amelia Earhart, to select the most impressive exhibits. A section of a spring, the propeller of an outboard motor and a self-aligning ball bearing took first, second and third place respectively.[17] Art lovers, even those with modern sensibilities, were not accustomed to seeing such things in museums, and the New York critics were snooty about the show. Undeterred, Johnson acquired a hundred of the exhibits as the core of what would become the museum's design collection.

Across the Atlantic, the German curator Alexander Dorner was experimenting with a different approach to exploring the relationship between art, design and architecture as director of the Landesmuseum in Hanover. Since the mid 1920s he had used pieces from its archive to depict the cultural history of particular eras by creating what he called 'Atmosphere Rooms'. The grand finale was the Raum der Gegenwart, or Room of Today, for which Dorner commissioned Moholy-Nagy to create an immersive sequence of images depicting glimpses of contemporary art, architecture, design, theatre and sport with screenings of experimental Soviet films including Sergei Eisenstein's *Battleship Potemkin* and Dziga Vertov's *Man with a Movie Camera*.[18]

Dorner also contributed to a radical programme of art and design exhibitions in London organized by Herbert Read at the London Gallery during the 1930s. The subjects ranged from Herbert Bayer's graphics to the work of leading Constructivist artists, including Moholy-Nagy. As a curator and critic, Read, who wrote the influential 1934 book *Art and Industry*, treated design with the same intellectual rigour as painting and sculpture, and commissioned eminent colleagues to contribute catalogue essays in that spirit. Dorner summed up Bayer's work in the catalogue for his 1937 show with: 'Combining photographs, typography and draughtsmanship in his new space and tension, he seizes the eye of the spectator and stimulates his imagination, developing influences, which have never been possible before. All this is only possible, both as a painter and a commercial artist, because we have here an active person who is really widening our world concept.'[19]

By the late 1930s, many of the Bauhaüslers who had fled Germany to escape Nazi oppression were founding art and design schools in their new countries, or teaching at them. Some of the most famous Bauhaus teachers took up powerful positions in the United States. Gropius and Breuer taught at Harvard, and Josef Albers settled at Yale, after he and Anni had taught at Black Mountain College in North Carolina. Mies and Moholy both ended up in Chicago. The former took a plum post as director of the architecture school at the Illinois Institute of Technology, while Moholy struggled to found two new progressive institutions, the New Bauhaus followed by the School of Design.

Once in the United States, both he and the Alberses remained true to the eclectic spirit of the mid 1920s Bauhaus. Sadly, Moholy died in 1946, too soon to realize his ambitions, although his experimental approach to design was sustained by his students and colleagues, notably György Kepes. The Alberses helped to create a laboratorial environment at Black Mountain, where artists like Robert Rauschenberg, Robert Motherwell and Elaine and Willem de Kooning worked alongside the composer John Cage and choreographer Merce Cunningham. Buckminster Fuller constructed the first geodesic dome there in 1948 as a summer-school project. Other prominent Bauhaüslers focused increasingly on their own disciplines over the years: graphics for Bayer; and architecture in the case of Gropius, Mies and Breuer.

Even so, the inclusive ethos of the Bauhaus may have proved more enduring had design not been characterized so definitively as a commercial medium after the Second World War. The die was cast by the tagline of *Time* magazine's 1949 cover story on Raymond Loewy: 'He streamlines the sales curve.'[20] And post-war perceptions of design were crystallized in Thomas Watson Jnr's aphorism 'good design is good business'.

There was nothing wrong with the phrase, not least because Watson's company IBM had proved that good design could, indeed, be good for business, after he succeeded his father as its president in 1952. Back then, Watson Jnr was a design agnostic, or so he claimed. The turning point came in 1955, when he received a letter from an IBM executive in the Netherlands saying: 'Tom, we're going into the electronic era and I think IBM's designs and architecture are really lousy.' Shortly afterwards, Watson Jnr visited the New York showroom of IBM's Italian rival, Olivetti, and was so struck by the contrast between its sleekly modern interior and the mahogany-panelled walls of his own company's offices that he went to see the Italian company's Milan headquarters. Realizing that his colleague was right, he recruited a fellow Second World War fighter pilot, the architect and industrial designer Eliot Noyes, and gave him a brief to modernize design at IBM.[21]

In 1956, Noyes commissioned the graphic designer Paul Rand to develop a new corporate identity: the stripy I, B and M letters that have symbolized IBM ever since. He also charged Charles and Ray Eames with enthusing the public about science, technology and mathematics by designing exhibitions for IBM, including a pavilion at the 1964 New York World's Fair, and a series of films.[22] IBM emerged as a role model of an enlightened post-war corporation. Some of its design projects were of such exceptional quality that they achieved far more than their immediate commercial objectives. Rand's corporate identity is a formally beautiful piece of typography, and the Eameses' films were remarkable exercises in design research and communication. Their 1977 film *Powers of Ten*, which seeks to demonstrate the power of a single number and the relative size of the universe in less than ten minutes, is still praised for its clarity by scientists and mathematicians as well as by fellow designers.[23]

Yet those five words from Thomas Watson Jnr continued to define perceptions not only of IBM's approach to design, but design in general, by reinforcing its stereotype as a commercial tool, whose purpose, spirit and impact were very different from the purity and expressiveness of art. Hence the assumption that art = good, design = not so good. Though, as the Eameses had demonstrated in their work for IBM, there were exceptions: designers who, in one way or another, had broken free of the commercial constraints that bound so many of their peers and treated design as a medium of intellectual enquiry and self-expression, much as modernist artists did art.

Moholy-Nagy set a dazzling precedent in the first half of the twentieth century by putting his Constructivist principles into practice by pursuing whichever challenges engaged him in a seamless fusion of art, design, science and technology. 'There is no hierarchy of the arts, painting, photography, music, poetry, sculpture, architecture, nor of any other fields such as industrial design,' he wrote. 'They are equally valid departures toward the fusion of function and content in "design".'[24] His experiments have had an enduring influence through Kepes's pioneering work on the construction of digital imagery[25] and were introduced to the mass market not only by Saul Bass's film titles, but those of one of their Chicago students, Robert Brownjohn, who devised thrilling opening sequences for the 1960s James Bond movies *From Russia with Love* and *Goldfinger*. Both films begin with Moholyesque montages of images. For *Goldfinger*, Brownjohn projected scenes from the movie on to a woman's gold-painted naked body, with a golf ball sliding down her cleavage, and Sean Connery's Bond running across her thighs.[26]

Among the stars of Italy's post-war design scene, the Castiglioni brothers – Achille, Pier Giacomo and, initially, Livio – developed a succession of thoughtful, humorous and elegant products for mass manufacture from their studio in Milan.[27] They used a tractor seat to make one chair and a bicycle seat for another. Eighty miles away in Turin, Carlo Mollino worked with the local artisans to produce deeply idiosyncratic furniture and interiors that reflected his love of the baroque, art nouveau, futurism, modernism, biomorphism, surrealism and the vernacular architecture of the desolate Aosta Valley high up in the Alps, as well as his passions for sex, skiing, cars, aircraft, cinema and the occult. Unlike the Castiglionis, who were willing to fulfil the commercial demands of their clients, albeit with work of such quality that it transcended them, the privately wealthy Mollino had the freedom to work solely on his own terms, and never treated design as anything other than a medium of self-expression.[28]

All four of them, together with Ettore Sottsass, Alessandro Mendini, Joe Colombo and Enzo Mari, conformed to the neo-Constructivist principles of Bruno Munari, who believed that the challenge of designing for daily life was too important to be relegated to a commercial role, and should be imbued with the values of art. Munari aired his views in his regular columns for the Italian daily newspaper *Il Giorno*, and published a collection of them in his 1966 book *Design as Art*. By then, the pop art movement was engaging with design as part of consumer culture, but the art world's attitude to it remained deeply ambivalent. For every artist who was critically engaged by design, such as Richard Hamilton, and Andy Warhol, to whom Yves Saint Laurent dedicated his pop art haute couture collection,[29] there were others who were deeply sceptical about it, like the car-crushing American sculptor John Chamberlain. During his 1971 retrospective at the Guggenheim Museum in New York, he exhibited an expensive sofa, designed by Verner Panton, in faux-reverential style in the lobby to parody what he saw as the pretentiousness of contemporary furniture design.[30]

Chamberlain's cynicism was shared by an emerging group of artists who rejected consumerism and were fiercely critical of what they saw as the shallow optimism of pop culture. In the forefront were Alighiero Boetti, Lucio Fontana, Michelangelo Pistoletto and other Italian members of the *arte povera* movement, who symbolized their contempt for the art market by using found or cheap materials in deliberately fragile, sometimes perishable work. Their ideals were shared by avant-garde architecture groups such as Archigram in Britain, Ant Farm in the United States and Italy's Archizoom and Superstudio, as well as by designers like Mendini, who translated the *arte povera* spirit into his 1974 Lassù project. Having mounted a simple wooden chair, like one a child might draw, on a pedestal, he poured petrol over it and set it alight on a derelict industrial site.[31] Mendini recorded the chair's destruction on film and in photographs: the work was the imagery, not the object.

Together with Archizoom and Superstudio, he was part of the rapidly expanding postmodernist movement, which critiqued what it saw as the intellectual turgidity of modernism by embracing its taboos, including kitsch and nihilism.[32] By the late 1970s, Mendini was producing flamboyantly stylized furniture whose role was primarily conceptual rather than functional or commercial, and had founded a new design group with Sottsass, Studio Alchimia. For his 1978 Redesign series of chairs, he created pastiches of classical modernist pieces, including a replica of the Wassily chair developed by Marcel Breuer at the Bauhaus in the mid 1920s with 'clouds' hovering above the back.[33] Many of Alchimia's defining themes were introduced to the mass market in a blaze of media coverage during the early 1980s by Sottsass and the younger designers with whom he formed the Memphis group. To mark its launch in 1981, Sottsass posed with his young collaborators for a team portrait in the Tawaraya, a 'conversation pit' built in the shape of a pastel-coloured boxing ring by the Japanese designer Masanori Umeda.[34] Eventually, a Tawaraya took pride of place in the Monaco home of the fashion designer Karl Lagerfeld, which was furnished mostly with Memphis pieces.

Like Lagerfeld's early fashion collections for Chanel, Memphis furniture reflected the postmodernist sense of irony, pastiche and theatricality which was evident in the work of artists like Julian Schnabel, David Salle, Robert Longo, Jean-Michel Basquiat and Cindy Sherman. So too did graphic design, starting with the cover of one of the most important early post-modernist texts, *Learning from Las Vegas*, the 1972 book by Robert Venturi, Denise Scott Brown and Steven Izenour which was designed by Muriel Cooper before her Damascene conversion from print to computing.[35] By the early 1980s, postmodernist graphics were introduced to a mainstream audience in Neville Brody's artwork for magazines like *The Face* and the formally elegant, richly symbolic record covers designed by Peter Saville for the Manchester bands Joy Division and New Order, which came to mean as much to some of the people who owned them as the music.[36] As a teenager in 1984, the artist Wolfgang Tillmans found 'a strange record with undecipherable digital type on a light pinkish grey' in the bargain bin of a Hamburg record shop. It was New Order's single 'Confusion', which had been released the previous year with a sleeve designed by Saville. After taping the record to listen to it on his Walkman, Tillmans displayed it 'as a "work of art"' in his bedroom. 'It was the obscurity of Peter's design, the fact that it didn't do what design was expected to do, that drew my attention,' he explained. 'And sitting on display in my room, it continued to hold my fascination. It took on a life on its own, in a co-existence with New Order's broken-up sound.'[37]

Even so, public perceptions of design were still dominated by its 'expected' commercial role, so much so that the word 'design' was omitted from both the 1975 first edition and 1983 revision of *Keywords: A Vocabulary of Culture and Society*, written by the influential British political scientist Raymond Williams. Despite billing itself as 'neither a defining dictionary nor a specialist glossary', *Keywords* is a reliable guide to the concepts which were considered to be culturally significant during the 1970s and early 1980s. When Williams published an expanded edition in 1983, he deemed it necessary to add words like 'anarchism', 'anthropology', 'ecology', 'ethnic', 'liberation', ‘sex’, ‘technology’ and ‘under-privileged’, but not ‘design’.[38]

By the turn of the twenty-first century, design was a central concern for the artists in the Relational Aesthetics movement identified by the French curator Nicolas Bourriaud. Maurizio Cattelan, Olafur Eliasson, Liam Gillick, Dominique Gonzalez-Foerster, Carsten Höller, Pierre Huyghe, Rirkrit Tiravanija and other members of the group approached art as a social exercise in which other people could participate, rather than as a finished work produced privately and controlled by the artist. Many of the issues addressed in their projects, including the functional environments devised by Tiravanija, embraced elements of design and its impact on our lives.[39]

Other artists have followed suit, either by exploring the legacy of twentieth-century modernist designers, as Nairy Baghramian, Ryan Gander and Simon Starling have done, or by interrogating the impact of mass-market design on contemporary life, like Isa Genzken, Christoph Büchel, Ai Weiwei and Mark Leckey. Often powerful and moving, their work has helped to foster a more nuanced understanding of design history, and stimulated debate on current design issues. Ai Weiwei, in particular, made a useful contribution to the long-running discussion on the changing definition of design in his role as artistic co-director of the 2011 Gwangju Design Biennale in South Korea.[40] The debate continued the following year in the thirteenth edition of the Documenta exhibition in the German city of Kassel, whose artistic director, Carolyn Christov-Bakargiev, questioned the continuing relevance of the concepts of ‘art’ and of the ‘artist’, suggesting that they may be reaching the end of their historical legitimacy and could be replaced by something different.[41]

All of this would have augured well for a constructive redefinition of the relationship between art and design, if not for an irritating distraction that distorted the art world’s perception of design almost as effectively as ‘good design is good business’ had done half a century before: ‘design-art’.

Like Watson Jnr’s mantra, design-art was a commercial phenomenon. The term was invented in 1999 by Alexander Payne, director of design at the auction house Phillips, to describe the contemporary design lots in a sale.[42] His intention was clear: to persuade art collectors that design, particularly recently designed furniture, had many of the same qualities as the paintings and sculpture they were accustomed to buying at auction. Financially, it made perfect sense, especially for an auction house like Phillips. Up until then, design collectors had focused on furniture by twentieth-century modernists such as Jean Prouvé, Le Corbusier and Charlotte Perriand, with the exception of a small cadre of devotees who bought new work by contemporary designers. All of that changed in the early 2000s, when art collectors began buying contemporary furniture. Commercial art galleries, such as Gagosian in New York and Franco Noero in Turin, started to represent designers alongside artists, and specialist contemporary design galleries opened in cities all over the world.

By 'design', most of them meant furniture, some of which was genuinely interesting – original and accomplished, whether formally, conceptually or technically – while other pieces were impractical and bombastic. As the art market soared, so did the price of 'design-art', but when recession struck after the 2008 banking crisis, many *ingénu* design collectors disappeared and the market promptly collapsed. The nadir was a sale at Phillips in New York in May 2010 of the art and design collection acquired by the American technology entrepreneur Halsey Minor in the heady days before he went bankrupt. All of the lots were being auctioned by his creditors, including one of the Lockheed Lounge aluminium chaises longues made by the Australian designer Marc Newson shortly after leaving art school in Sydney in the late 1980s. Two years before, it had set a record for a work by a living furniture designer when Minor bought it privately for $2.25 million. This time, it raised just $2 million while two other pieces by Newson were unsold, and a fourth went to a lone bidder, reportedly his gallerist Larry Gagosian.[43]

Rollercoaster prices notwithstanding, the principal problem with design-art was that by dominating the art world's perceptions of design, it came to be seen as the area of design which was closest to art, not least as many of its practitioners spoke at length about the 'sculptural' and 'artistic' qualities of their work. Commercially this was the case, but not culturally. Design-art may have been bought and sold like paintings and sculpture, but much of it was of questionable cultural value and not nearly as formally resolved, technologically innovative or intellectually provocative as thoughtful examples of industrial design. There is a similar tension between commercial and critical values in the art world, where the most expensive works are not necessarily those of the greatest cultural import. But the market for art is so much larger and more mature than it is for design-art that the nuances are more apparent. Design-art also seemed so ubiquitous that it obscured other arguably deeper, more demanding areas of design, thereby perpetuating the misperception of design as a shallow medium with none of the rigour or complexity of critically engaging art. With luck, the design-art bubble will turn out to have been a fleeting diversion, because during its rise and fall other areas of design had changed dramatically, particularly with regard to their intellectual weight and nuance.

Traditionally, an important distinction between art and design was that artists made their work themselves, whereas designers delegated production to other people, generally to specialists in particular industrial or artisanal processes. Artists had deviated from this stereotype since Marcel Duchamp's early twentieth-century experiments with industrial production.[44] By the 1960s, Donald Judd, Barnett Newman and Sol LeWitt regularly outsourced the making of their sculpture to specialist workshops, such as Treitel-Gratz Co in Manhattan, which was previously best known as the American manufacturer of Mies van der Rohe's Barcelona chair.

By the early 2000s, more and more artists were outsourcing production. Some followed Judd, Newman and LeWitt by using the same specialist manufacturers as designers, as the American sculptor Richard Serra did by working with Bethlehem Steel in the United States and later Pickhan in Germany. Others turned to the rapidly expanding cottage industry of art fabricators, including Carlson & Co in San Fernando, California, and the Mike Smith Studio in London. As well as making all or part of artists' work, these workshops helped them to research and develop specialized materials and technologies. Some artists, including Olafur Eliasson, transformed their studios into multidisciplinary research and development units staffed by architects, engineers and computer programmers as well as artists. Fabrication became so prevalent that *Artforum* magazine devoted a special issue to it in October 2007 entitled 'The Art of Production'.

At the same time, digital technology was enabling designers to exercise closer control over the development of their work and, if they wished, its production. Traditionally, industrial designers had sent sketches and technical drawings of their products to manufacturers, who then made models based on their specifications before engineering prototypes for production. As computer-aided design software advanced, they were able to take charge of the development process up until the final stage of prototyping, thereby reducing the risk of deviations being introduced during modelling and prototyping. At least, this is the theory of what might happen. In practice, the designer's intentions could still be compromised by the manufacturers' demands, which weaker designers gave in to. But, in principle, designers have been able to develop work which is more expressive and complex, in terms of its form and the messages it communicates. There is a parallel with film-making in that digital technology has empowered the designer to become, not an artist, but what the director and critic François Truffaut defined in the 1950s as an auteur, by ensuring that the finished piece reflects his or her vision.[45]

The auteur phenomenon has been most dynamic in graphic design. Back in 1986, the American designer April Greiman was asked to design a poster for the Walker Art Center in Minneapolis. She decided to produce a life-size nude self-portrait on her Macintosh computer, which had been introduced two years before, but the process turned out to be tougher than she had expected. When Greiman finally printed the image, the files absorbed so much of her computer's memory that the laser printer nearly died. Those printer-crushing files were two hundred and eighty-nine kilobytes in size, just big enough to store a fuzzy photograph snapped by a phone.[46] Intrepid though the exercise seemed at the time, it is now the type of thing that anyone (even those of us who are neither graphic designers nor programmers) can do on a cheap computer.

Graphic designers have since broadened their roles to become authors, producers, publishers, curators, bloggers and entrepreneurs. The British critic and curator Rick Poynor explored the impact of digital technology on the graphic design process and its role in encouraging designers to produce more self-expressive work in his 2003 book *No More Rules: Graphic Design and Postmodernism*.[47] And the Walker Art Center analysed the role of social media, mobile devices, print-on-demand systems, rapid prototyping and web-based distribution in enabling them to experiment with new forms of production and distribution, and to develop their own design tools, in a 2011 exhibition 'Graphic Design: Now in Production'.[48]

One consequence is that graphic design has become more formally sophisticated, in terms of both its visual appeal and its structural complexity. The designer of this book, Irma Boom, has produced a succession of remarkable books in which she has combined different scents, textures, hidden motifs, colour codes, unusual bindings and papers with typefaces that become progressively smaller – or larger – as the text continues. Having printed one book on coffee filter paper to achieve the desired effect, she hacked at the edges of another with a circular saw. Designing and editing a particularly ambitious book, a history of the Dutch conglomerate SHV, absorbed five years of Boom's working life. The page edges are designed to reveal a line of poetry from one angle, and an image of tulips from another. The title only becomes visible on the cover after regular use. There are two thousand, one hundred and thirty-six pages in the book, but Boom insisted on publishing it without page numbers or an index to encourage readers to dip in and out, rather than to read it in sequence.[49]

Other graphic designers have used their new-found freedom to treat the design process as a medium of intellectual enquiry. Daniel van der Velden and Vinca Kruk, co-founders of Metahaven in Amsterdam, have dubbed this practice 'speculative design'. Rather than waiting for commercial clients to set a brief, they identify an issue that intrigues them, such as a political or economic phenomenon, and conduct an unsolicited design exercise, for example by taking it upon themselves to design a visual identity for a country or new form of currency. Dismissive though the Metahaven duo are of the intellectual limitations of commercial design practice, they use its processes as their primary tools and adhere to its rituals in their work.[50]

A similar approach is evident in other fields. The British designers Anthony Dunne and Fiona Raby have played decisive roles in the evolution of what they dubbed 'critical design' through their own practice and as teachers at the Royal College of Art in London.[51] Much of their work is devoted to developing conceptual design projects which critique design and consumer culture, thereby occupying similar terrain to conceptual artists and cultural historians. Often they address issues that have long been taboo within design, including negative emotions, such as our fears and neuroses. In one project with the self-explanatory title 'Designs for Fragile Personalities in Anxious Times', Dunne and Raby designed pieces of furniture that are intended as defences against such horrors as alien invasion and nuclear disasters.[52]

Other designers have applied the principles of critical design in a commercial context, as Martí Guixé did by producing a series of carrier bags for the Spanish shoe retailer Camper bearing the handwritten slogan 'Don't buy it if you don't need it'.[53] More often, they are applied on a smaller scale to conceptual projects intended as provocations rather than destined for mass production. The German designer Julia Lohmann modelled her Cow benches on the shapes of particular cows' backs and upholstered them in cow hide, thereby making them impossible to sit on – or even look at –without thinking of the animals they were made from. Each bench has been given a name, as if it were a pet, such as 'Waltraud', named after Lohmann's mother, thereby illustrating our hypocrisy in pampering individual animals, yet slaughtering anonymous cattle en masse.[54] Our ambiguity towards animal politics is also the theme of a project by the Dutch designer Christien Meindertsma, who produced a book, *Pig 05049*, in which she traces all the products made from a single pig: cigarettes, heart valves, ice cream, injectable collagen, crayons, beer, bullets, antifreeze, cellular concrete, a tambourine, chewing gum, toothpaste and even the glue used to bind the book as well as steaks, chops, ribs and sausages. While the French designer Mathieu Lehanneur has analysed the insecurities and inconsistencies of human nature in his work in health care.[55]

Not only are the processes of producing art and design aligning, but by casting themselves as auteurs and activists, designers are exercising their right to use their work as a medium of self-expression and research, as artists have traditionally done, free from the restrictions imposed by the demands of commercial design briefs. Does this mean that the end result is the same as art, or that it should be defined as art rather than design?

I would say no on both counts. Firstly, design always has a designated function, regardless of whether it is determined by the commercial objectives of a corporate client or by the designer's intellectual curiosity. Nor does that function necessarily need to have a practical purpose or to be of commercial value; it could be to communicate a political message, or to enable the designer to embark on a research exercise. Deconstructing the social, cultural and political identity of a nation may seem very different to finessing a corporate symbol for a multinational like IBM, but each exercise fulfils the required function. Works of art can, of course, be functional too, but need not necessarily be so, unlike design: a critical distinction between the disciplines. Remember Donald Judd: 'If a chair . . . is not functional, if it appears to be only art, it is ridiculous.'[56]

Another distinguishing factor is that every design project – commercial, conceptual or critical – is defined by design culture, at least to some degree. Perhaps the design process was applied to its development, or the finished work incorporates design techniques and design references. Alternatively, it might explore an aspect of design's history or its impact on contemporary life. Again, works of art can do the same, but they can also choose not to do so, which gives them a freedom and intensity of expression that design is denied. Anthony Dunne has likened the relationship between art and design to that of science and engineering, with the former being devoted to pure research and the latter to applied research, though in design's case that research will always be applied to daily life.[57]

Thanks to his and Fiona Raby's work, as well as that of Ai Weiwei, Metahaven, April Greiman, Julia Lohmann, Christien Meindertsma, Mathieu Lehanneur, Alessandro Mendini, Bruno Munari, György Kepes, László Moholy-Nagy and all the other practitioners who have pushed against the traditional boundaries of the field, it is now possible for designers to fulfil many of the roles of artists without needing to define the outcome as art. Even so, many of the most complex and eloquent examples of design are entirely conventional in concept and execution, yet can still play Mont Sainte-Victoire to a contemporary artist's Cézanne, as Dieter Rams's mass-manufactured electronic products for Braun once did to Richard Hamilton.

Of all the many thousands of words devoted to untangling the tortuous relationship between art and design, my favourites are those of Charles Eames, who, when asked if design was 'an expression of art', said: 'I would rather say that it is an expression of purpose. It may (if it is good enough) later be judged as art.'[58]

8 Sign of the times

I had dinner here in Palm Springs a few years ago with Mr and Mrs Donald Salem. A dinner party. Black tie – ridiculous in Palm Springs. My neighbour at my table was a lovely young lady, and she asked me suddenly: 'Why did you put two xxs on Exxon?' I asked her: 'Why ask me?' She said, 'Because I couldn't help seeing it.' I replied, 'Well, that's the answer.'
— Raymond Loewy[1]

Take two corporate logos. Each one belongs to a company whose name begins with 'A'. Neither name bears any relation to what the business does. In each case, their chosen symbol is an image of an apple with a leaf sprouting from the top and a bite nibbled out of one side. But there the similarities end, because everything else about those seemingly identical logos is different, as are the companies they belong to.

One of them is Apple; and the other is Ann Summers, which, far from being a global technology group, is Britain's biggest chain of sex shops. By any definition, they are an odd couple, which should make it all the more surprising that they have adopted such similar visual identities, especially as each of them has done so successfully. The golden rule of corporate identity design is that the result should reflect the spirit of the company by symbolizing what it stands for, how it behaves and what it can realistically hope to achieve. Both Apple's and Ann Summers' identities do. How can what sounds like more or less the same motif symbolize two such diverse organizations, yet do so accurately?

The explanation is that, although the components of the two identities are similar, the ways in which their respective designers have executed them are not. The choice of colour, shape, typography and exactly what the apple alludes to are very different in each case. And it is those design details that send us the visual clues with which we can work out what the symbol is trying to say.

First, Apple. Steve Jobs and Steve 'Woz' Wozniak picked

the company's name – originally Apple Computers – when they founded the business in 1976. Not only was Jobs fond of fruitarian diets at the time, he had recently returned from pruning the apple trees at the All One Farm, a commune near Portland, Oregon, that he often visited at weekends. There were rational reasons for the choice of name too. 'It sounded fun, spirited and not intimidating,' Jobs explained, years later. 'Apple took the edge off the word "computer". Plus, it would get us ahead of Atari in the phone book.'[2]

The original logo was drawn by Ron Wayne, who was helping to set up the business, as an illustration of the seventeenth-century scientist Isaac Newton hatching the theory of gravity after seeing an apple fall from a tree. He drew it in the ornate style of Victorian children's books and the artwork of their favourite late 1960s Bay Area rock bands. The following year, that logo was replaced by a simpler symbol devised by a local graphic designer Rob Janoff in the shape of an apple. He produced two versions: one of a whole fruit; and the other of one with a bite taken out of the side. Jobs picked the latter on the grounds that the unbitten apple looked too much like a cherry,[3] although a favourite theory of Apple mythologists is that the bite alludes to a computer byte. (Equally popular is the myth that the symbol is a tribute not only to Newton's theory of gravity, but to the death of Alan Turing, the gifted British mathematician who was at the forefront of computer science from the mid 1930s until his suicide in 1954. Turing killed himself after being convicted of 'gross indecency' with another man and subjected to chemical castration. A half-eaten apple was found near his corpse, fuelling suspicions that he had laced it with a deadly dose of cyanide.[4]) As a finishing touch to Apple's motif, Janoff added stripes in the colours of the rainbow as a nod to the rainbow flag flown by Greenpeace and other counter-culture groups, as well as a cheeky pastiche of the monochrome stripes in Paul Rand's IBM logo.

Apple's emblem has since lost the stripes, but kept roughly the same shape of apple, rendered as a solid block of colour, generally in shades of grey, black, white or silver. The company has been equally particular in its choice of typefaces. When the Macintosh computer was introduced in 1984, Apple adopted its own version of Garamond, an elegant sixteenth-century serif font. In 2002 it switched to a sans serif typeface, but chose an unusually curvaceous one, a customized version of the digital font Adobe Myriad.[5]

What do we see when we look at Apple's identity? Something that is elegant, confident and disciplined: sophisticated enough to appreciate the work of a great scientist like Isaac Newton, but not too serious to baulk at naming itself after a fruit. Even so, the logo also suggests that Apple is no longer quite so subversive that it wants to be associated with political activism or to be seen as a brattish upstart poking fun at a corporate titan like IBM.

And what does Ann Summers' identity say? The name is an abbreviation of Annice Summers, who was the secretary of the company's founder, the former actor Michael Caborn-Waterfield, when he began the business in 1970. She left soon afterwards, and he sold out to the brothers Ralph and David Gold in 1972. Their strategy was to introduce sex shops to the mass market by locating them on busy shopping streets rather than back streets.[6] If they were to succeed, it was important that their new customers, especially women, didn't feel embarrassed when entering the stores, and they kept the name Ann Summers because it sounded suitably warm, friendly and familiar. Ms Everywoman, in fact.

The Golds also needed their stores to seem sexy and fun, which is where their take on the apple motif came in. Rather than alluding to scientific history, their company's apple refers to the forbidden fruit that Adam and Eve failed to resist in the Garden of Eden. Whereas Apple's symbol is round, Ann Summers' is shaped like a heart to evoke love. The leaf resembles a flame to signify the heat of passion, and the bite represents the consequences of succumbing to temptation. Even the typeface is playful in style, with flirtatious touches, such as a diagonal stroke in the centre of the 'e'. And the colours of the brand name emblazoned above each Ann Summers store are gaudy shades of pink and red, to symbolize flesh and sex respectively.[7]

In short, Ann Summers' corporate symbol is as different from Apple's as the story of Adam and Eve's fall from grace is from Isaac Newton's discovery of the theory of gravity. When we see those two symbols, we understand the distinction instinctively by decoding the visual clues they contain. People have responded to signs and symbols in the same intuitive way throughout history. All the visual cues they have picked up on were designed in the sense that they were conceived and executed with the intention of conveying a specific message, even though it is only relatively recently that professional

Tomáš Gabzdil Libertíny's
Honeycomb Vase made
by thousands of bees

Cow benches, designed
by Julia Lohmann

The *Pig 05049* book, designed by Christien Meindertsma

A replica of an illegal computer recycling plant in Christoph Büchel's 2006 installation, *Simply Botiful*

designers, like the ones responsible for Apple's and Ann Summers' identities, have been paid to produce them.

One of the richest sources of intuitively designed symbols is political activism. The raised fist has been an emblem of protest for a succession of radical movements, from the 1917 Russian revolutionaries and anti-fascist insurgents in the Spanish Civil War during the 1930s, to the Black Power movement in the 1960s and recently for some of the Occupy anti-capitalist groups. But its origins date back to ancient Assyria, when Ishtar, the goddess of love, war, sex and fertility, was depicted raising a clenched fist to symbolize strength in the face of violence. However often we have seen that emblem in different guises since then, we still recognize what it is saying because it was so well chosen – the physical gesture of clenching and lifting your own fist really does make you feel stronger – that it has become inseparable from its original meaning.

The same can be said of one of the oldest peace motifs, the olive branch. It derives from ancient Greek mythology, when the gods Poseidon and Athene were wrangling over the city of Athens. No sooner had Athene contested Poseidon's claim to own the area around the city by planting an olive tree there, than he furiously challenged her to one-on-one combat. Zeus intervened and insisted that their dispute be settled by arbitration. All of the gods and goddesses were summoned to vote. Zeus abstained, but the other gods voted for Poseidon and all the goddesses for Athene. As the sexes were represented equally, the gods were outnumbered by dint of Zeus's abstention, Athene won by a single vote, and Athens became hers.[8] To this day, the phrase 'passing the olive branch' is used to describe an attempt at peacemaking.

New symbols have since been coined to represent different needs, desires, ideals and fears. Some of them are instructive, and are intended to regulate our behaviour or to tell us to perform particular tasks. The decision of the sixteenth-century Welsh mathematician Robert Recorde to replace the words 'is equal to' with two parallel horizontal lines of identical length, thereby inventing the equals sign, '=', is an exceptionally successful example of designing an instructive symbol. In his 1557 book *The Whetstone of Witte*, Recorde recalled having become so bored by the 'tedious repetition' of writing those words that he decided to abbreviate them to '=' on the grounds that 'no two things can be more equal' than those lines.[9] Not that

Recorde's brainwave was recognized for its design merits at the time – though neither was the invention of another instructive symbol when Edward 'Blackbeard' Teach, Bartholomew 'Black Bart' Roberts and other early eighteenth-century pirates added diabolic skulls and crossed bones to their flags.[10]

Throughout the twentieth century, one of the most important roles of designers was to help people to understand what was happening in the world around them by devising signs and symbols that conveyed useful information and offered guidance, alerting them to possible hazards, or showing them how to find their way from A to B. Think of someone who is taking their first flight and trying to work out where to go in the labyrinthine environment of an airport, an artificial space, much of which may be devoid of natural light and where the consequences of not being in the right place at the right time can be dire. How could they cope without well-designed signs, like Ruedi Rüegg's at Zurich Airport? The same applies to trying to navigate the tangled mass of tunnels in a subway station, where you are below ground level with no means of identifying your location other than the signs, which you may need to read at speed, while distracted by crowds of passengers and garbled announcements, as New Yorkers learnt to their cost before their subway signage was rationalized.

An impressive example of this form of instructive symbolism is the British road-sign system designed by Jock Kinneir and Margaret Calvert in the late 1950s and early 1960s. Until then, most roads in Britain had sported a motley assortment of signs in different styles, sizes, formats and states of disrepair, which had accumulated over the years with the old ones remaining, however dilapidated they had become, after new versions were added. The result was confusing, perilously so for motorists, who would become accustomed to deciphering one type of sign only to have to waste valuable time puzzling over another, and risk being distracted from what was happening on the road. The graphic designer Herbert Spencer illustrated how chaotic the roads had become by photographing each of the hundreds of signs he came across when driving along the A3 from central London to the recently opened Heathrow Airport in 1961, and publishing the results in consecutive issues of *Typographica* magazine.[11]

By then, Kinneir and Calvert had been commissioned to design a signage system for Britain's newly built motorways, but they owed their appointment to the type of nepotistic coincidence that often determines such decisions. The chairman of the government committee responsible for motorway signs was Colin Anderson, who also chaired the shipping company P&O-Orient Lines for which Kinneir had devised a brilliantly simple labelling system that helped the staff to track the passengers' luggage as it was shuttled between ships.[12] Other European countries were constructing motorways at the time, and typically left signage design to the engineers, who tended to treat it as a peripheral project. Britain may well have done the same, if Anderson had not remembered how effective Jock Kinneir had been at P&O.

He and Calvert were charged with devising a coherent system of signs that could be understood quickly and easily by motorists wherever they were in the country and whatever the weather, even when driving at speed. Every element was rigorously tested to ensure that it was as clear and consistent as possible: the style and size of the lettering; whether the letters should be upper or lower case; the size of the spaces between them, and of the borders surrounding them; the colours; what the signs would be made of; how they should be supported; how high they should stand above the ground. The prototypes were tested at Hyde Park in London and on the newly built bypass at Preston in Lancashire. The results were so successful that when Herbert Spencer's exposé goaded the government into taking action to improve the country's road signs, Kinneir and Calvert were the obvious choices to design them too.

Both the motorway and the road signs needed to provide clear, coherent ways of using letters and numbers to indicate distances, speed limits and to tell drivers where to turn off on to other roads. This was why Kinneir and Calvert devoted so much time to ascertaining whether their choice of typography, the order of information and its positioning was as clear as possible. But the signs also required a means of alerting drivers to whatever they might find along the route. Whether they were approaching a bridge, bend, tunnel, steep slope, ford, one-way street, dead end, roadworks, junction or level crossing, there had to be a sign saying so, just as there needed to be signs to forewarn them of the risk of debris tumbling on to the road, or of cattle crossing it and bringing the traffic to a halt. Kinneir and Calvert decided that the most effective solution would be to produce a series of pictograms, each illustrating a particular feature of the roads or possible hazard.

Kinneir took charge of devising the overall system, and

Calvert of the pictograms. Stylistically, they needed to be as uncomplicated as possible to ensure that drivers could understand them instantly, even when squinting through torrential rain. They also had to be recognizable to people of different ages and backgrounds in different parts of the country. Calvert drew most of the symbols as silhouettes in solid simplified forms that a child might have sketched, depicting people's heads as circles, for instance. For the 'children crossing' sign, which was to be positioned near schools, she referenced her own childhood memories by basing it on a photograph of her leading her younger brother by the hand.[13]

The overriding objective was clarity, but she also imbued her pictorial symbols – and Kinneir's meticulously ordered system – with a human quality. Their signs fulfilled their designated function of telling British drivers exactly what to do and when to do it, but did so courteously. Many of Calvert's symbols remain intact today, more than half a century after she designed them. Some have inevitably been tweaked or replaced, but the results seem so unsatisfactory in comparison that they often act as forlorn reminders of the originals. The signage system that Margaret Calvert created with Jock Kinneir has been imitated all over the world, literally in the case of some countries, which have imported some elements intact. Many subsequent examples of inspired instructive signage, such as Benno Wissig's 1967 system for Schiphol Airport in Amsterdam,[14] Adrian Frutiger's sadly neglected scheme for Charles de Gaulle Airport in Paris and Ruedi Rüegg's signs at Zurich Airport, have owed something to their work together.

More recent instructive symbols have been the products of intelligent reinvention, rather than being designed from scratch. Take the @. No one knows for sure where it came from. One theory is that it was introduced in the sixth or seventh century as an abbreviation of *ad*, the Latin word for 'at'. Just like Robert Recorde when he abbreviated 'is equal to' by inventing the equals sign in 1557, the scribes of the day are said to have wearied of writing out both letters, and decided to combine them by curling the stroke of the 'd' around the 'a'. Another explanation is that it originated in sixteenth-century Venice, when it was adopted as shorthand for the amphora, a measuring device used by local tradesmen.

Whatever the truth, the @ came into its own in the late nineteenth century when it appeared on the keyboard of the first typewriter, the American Underwood, in 1885. It was then used as an abbreviation of the accounting term 'at the rate of'. Even though its use diminished over the years, the @ remained a typewriter key and, as a result, appeared on early computer keyboards. It is thanks to its inclusion there that the @ has enjoyed a second coming.

In 1971, the American computer programmer Raymond Tomlinson was preparing to send a message from one computer to another in the first email. He wrote the addresses of the sender and recipient in computer code, but needed to translate it into a form of wording that we 'civilians' could understand. Having decided that the first half of the address should identify the user, and the other half their computer, he looked for a means of indicating that he or she was 'at' that machine.[15] The @ seemed perfect. Not only was it convenient to use, by dint of being on the keyboard, its old meaning was reassuringly similar to the new one. At least it was to the few people who still used it. By the early 1970s the @ was used so seldom that it could embark on its new role as the symbolic equivalent of a comeback kid with very little emotional baggage from the past, in an inspired exercise of design thinking.

The same could be said of another recently reinvented instructive symbol, the #, known as the hashtag, pound sign or number key. (An attempt in the 1960s to rename it the 'octothorp' or 'octhorpe' failed.) Like the @, the # was an early fixture on computer keyboards, where it was chiefly used as an abbreviation for 'lb', or a pound in weight. But some manufacturers demoted it from the keyboard by making it one of the secondary symbols, which are created by pressing several keys simultaneously. The # did, however, fulfil a useful function on phone keypads, where it was often used, together with the star key, when callers performed remote functions, such as retrieving voicemail or making credit card payments.

The ubiquity of the # eventually won it a dynamic new role as a Twitter identification tag. On 23 August 2007, the San Francisco-based tech activist Chris Messina, known as FactoryJoe, sent a tweet asking: 'how do you feel about using # (pound) for groups. As in #barcamp [msg]?'[16] Eager to find a way of enabling himself and fellow tweeps to follow particular themes by logging on to a collection of all the tweets on those subjects, he suggested attaching # as a prefix – to, say, '#buckminsterfuller' or '#tsunami' – as a means of achieving this.[17]

The take-up was relatively slow until San Diego County in southern California was ravaged by wildfires in 2007. More than half a million people were forced to leave their homes, and thousands of buildings destroyed.[18] A local web developer, Nate Ritter, decided to track the disaster by posting regular updates on Twitter and used the hashtag '#sandiegofire' to flag them to anyone who wanted to know what was happening.[19] By providing sorely needed information to people in a perilous situation, his hashtagged tweets demonstrated how effective Messina's idea of reviving the # as an instructive symbol for social media could be.

Inspired though such motifs are – and the @'s reinvention has won it a place in the design collection of New York's Museum of Modern Art[20] – another type of emblem with a different role can be equally useful: the descriptive symbols intended to identify individuals or groups of people and to explain what they have in common. The raised fist and olive branch fall into this category, as do newer political symbols, such as the pink triangle, which became a popular motif for gay rights groups in a tribute to the homosexual prisoners in Nazi concentration camps, who were identified by it in the official records.[21] Equally eloquent is the circular anti-nuclear symbol devised by the British designer Gerald Holtom in 1958 for the Campaign for Nuclear Disarmament's protest march from Trafalgar Square in London to the Atomic Weapons Establishment at Aldermaston in Hampshire. It consists of a circle enclosing two diagonal lines and one vertical, which represent the semaphore signals for the letters 'N' in 'nuclear' and 'D' in 'disarmament'.[22]

A more formal genre of descriptive symbols dates back to the ancient heraldic motifs adopted by aristocratic warriors in the mid twelfth century for use in battle. Originally, these emblems were conceived as tactical ploys to identify individual soldiers in combat, such as those fighting for the English king Richard I in the Crusades, who had become difficult to distinguish once they took to wearing full armour. Their coats of arms – or armorial bearings, as they were called – were attached to their shields or breastplates where they were clearly visible, and afforded extra protection to the wearer. Soon, the same symbols were being worn by warriors' squires, and then by their foot soldiers, priests, clerks, tenants, servants, the labourers who tilled their land, and anyone else claiming to be associated with them. Eventually there were so many abuses of coats of arms, with people sporting heraldic motifs to which they had no rightful claim, that some countries introduced laws to regulate their use.

Cities, towns, villages, schools, universities, churches and sports teams have all adopted their equivalent of Crusaders' armorial bearings over the centuries, always with the same objective of identifying themselves and imbuing the people who have an allegiance to them with a sense of belonging. Businesses have done the same. Corporate identities, like Apple's and Ann Summers' fruit motifs, are among the commonest forms of descriptive symbols, so common that a typical Western consumer is said to be exposed to several thousand of them every day.[23] If you doubt that you see quite so many, just think of all the logos you spot, but do not necessarily notice, on food packaging in the fridge, trucks driving by, carrier bags crumpled in the gutter, advertising billboards, ticket stubs, websites, building signs, T-shirts, corporate stationery, email signature pickers and television commercials.

Many of the earliest corporate identities resembled aristocratic coats of arms, largely because fledgling companies wanted to reassure prospective customers that they were robust and reliable, rather than spivvy fly-by-nights who were likely to rip them off, or disappear after a few months. To that end, they sought to present themselves as being as powerful and enduring as the landed aristocracy. Pretentious though this sounds, it was a prudent strategy given that until the late nineteenth century most people bought goods from local shopkeepers, craftsmen or other suppliers whom they knew personally. Even tinkers and travelling salesmen tended to be familiar to their customers because they stuck to the same routes. Once businesses began shipping their goods further afield on newly built roads and railways, they needed to find other ways of convincing people who had never actually met them that they were reputable and trustworthy. One solution was to identify their products, trucks and paperwork by branding them with distinctive symbols. Often, the same companies also wanted to persuade their employees that they were powerful and imposing: qualities associated with the feudal warlords of past centuries. Traces of these faux heraldic emblems are still visible in some corporate logos, including BMW's and Fiat's, whose circular motifs look as though they could have come from warriors' shields. Grander still is the emblem of Santa Maria Novella, the Florentine perfumery, which adopts the same shape as a traditional coat of arms, topped by a crown, in a regal palette of royal blue and gold.[24]

Other companies have plumped for symbols that say something about themselves, by describing what they do or what they stand for, rather than claiming aristocratic authority.

A popular option was to try to replicate the personal nature of traditional trading by identifying the business with a signature, so that it looked as if someone, usually the founder, was endorsing its goods. Some of those signatures were genuine. When a young American broom salesman, Will Keith Kellogg, founded the Battle Creek Toasted Corn Flake Company in Battle Creek, Michigan, in 1906 to make breakfast cereal to his recipe, he printed his signature on every packet.[25] The Kellogg Company has used variations of it ever since. Similarly, the extrovert Australian industrialist Macpherson Robertson named his confectionery company MacRobertson in the 1880s, as an abbreviation of his own names, and chose a fancy version of his signature as its motif. His 'signature' was written in neon lights above his factory in Melbourne and became a local landmark.[26]

Some signature logos are fictitious and were created to give the impression of belonging to a particular person. The recipe for Coca-Cola was invented in 1886 by Dr John Pemberton, a wounded army veteran turned pharmacist in Atlanta, Georgia, who hoped to sell it as a headache cure. Once it went on sale at a local soda fountain, the company's book-keeper Frank Robinson suggested calling the cure Coca-Cola after two of its main ingredients, coca leaf extract and kola nuts, whose 'k' was replaced by a 'c' to make the name sound catchier. Robinson wrote it out in the then-fashionable ornate style of lettering known as Spencerian script.[27] Coca-Cola was a hit, not as a headache cure, but as a pick-me-up (probably thanks to the traces of cocaine in the coca leaf extract it then used, and to the caffeine in the kola nuts), and the Coca-Cola Company still uses the Spencerian-style signature as its logo.

Another approach was for companies to illustrate the things they made or did in 'biographical logos'. The French luxury good house Hermès was founded in 1837 by Thierry Hermès, the orphaned son of an innkeeper, who opened a work-shop in Paris to make horses' harnesses of such high quality that Europe's wealthiest families vied to buy them. These days, Hermès is better known for the very long waiting lists to buy its Kelly and Birkin bags, but it still makes harnesses in the tiny saddlery tucked above its flagship store on rue Faubourg Saint-Honoré in Paris, and its corporate symbol is an homage to Thierry Hermès's trade: depicting a man in a top hat and breeches standing beside a smartly harnessed horse pulling a nineteenth-century open-top carriage.[28] The red and yellow logo of my favourite football club, Manchester United, looks like a heraldic coat of arms at first glance, but on closer inspection sports a couple of footballs and a perky devil brandishing a trident in a nod to the 'Red Devils', the nickname given to the team in the early 1960s.[29] Originally the name belonged to a local rugby team, but when Manchester United's manager Matt Busby heard it, he suggested that the club should adopt it as a suitably feisty name, which was likely to unsettle opposing teams. The new nickname became so popular that the red devil motif was added to the club's official visual identity.

Similarly, the logo of Citroën, the French car company, sports a pair of upturned chevrons in a nod to the ingenious gear-cutting process that its founder, André Citroën, discovered on a trip to Poland in 1900. He began the business by using that process to produce gears in the shape of chevrons.[30] Decades later, NASA, the National Aeronautics and Space Administration, introduced a biographical logo in 1959, officially known as the 'NASA insignia' and unofficially as the 'meatball'. Designed by an employee, James Modarelli, it consists of a spacecraft orbiting the letters N, A, S and A in a starlit sky (the stars really do look like the dimples of a meatball) beside the elongated V shape of the latest hypersonic wings. As well as speaking volumes about the NASA team's passion for exploring outer space, it alludes to the amateur design tradition there. It is customary for a member of the crew for each mission to design a commemorative patch for the entire team to wear on their spacesuits. The meatball was nudged into retirement in 1972 when NASA adopted a new logo, the futuristic 'worm', as part of the Federal Design Improvement Program implemented by the Nixon regime. But when NASA was restructured in 1992, after a traumatic period in the 1980s, the meatball was reinstated as a rousing symbol of the good old days of the space race.[31]

Fond though space nuts are of the meatball, design purists may have considered it dated of NASA to have commissioned a biographical motif in the late 1950s. By then, businesses no longer wished to be associated with one person or a particular product as if they were traditional mom 'n' pop shops. In an era when multinational conglomerates were diversifying into new industries and opening offices all over the world, corporate success was equated with speed and scale. Go-getting post-war corporations sought descriptive symbols that conveyed an aura of dynamism and sleek inscrutability, rather like the mid-century modern skyscrapers, which many of them had commissioned as their headquarters. This was the heyday of the anonymous corporate symbol, often consisting solely of the company's name – or initials – spelt out in a specific typeface, like Paul Rand's I, B and M emblem. A well-designed corporate identity was regarded as one that was both distinctive and memorable. At the time, the consensus was that it should look exactly the same wherever it appeared, which Rand achieved brilliantly in his masterly handling of IBM's symbol.

First, Rand did not start off with stripy initials, but restricted himself to selecting a more modern and refined typeface to spell out I, B and M. As Steven Heller explained in his monograph of Rand's work, he waited until he had won his client's confidence before suggesting the stripes. As well as making IBM's otherwise rather mismatched initials look more coherent, he argued that the stripes gave those letters an authoritative air by evoking the skinny parallel lines that act as anti-forgery devices on the signature sections of legal documents.

Second, Rand realized that ensuring the identity was executed to the same high standard wherever it was used would be just as important as what it looked like, and probably trickier to pull off. Like many post-war multinationals, IBM owned factories, warehouses, offices and showrooms all over the world, and the implementation of its identity would be delegated to different designers, draughtsmen and printers in each location. Printing standards varied from place to place, and many of the designers were accustomed to doing more or less what they wanted. One enterprising employee at IBM's production plant in Poughkeepsie, New York, had invented a cartoon character, Ogiwamba, who featured prominently on the factory posters. Inevitably, he and his peers were not best pleased at being told to jettison their ideas and to obey Rand's edicts. Undeterred, Rand held regular briefing sessions with groups of IBM designers at its Manhattan headquarters and summoned them to his home.[32] Sometimes they repaired for informal meetings to his favourite local restaurant, Gold's Delicatessen in Weston, Connecticut.[33] Rand wrote two lengthy documents – 'IBM Logo Use and Abuse' and the less ominously entitled 'The IBM Logo' – in which he specified how the identity should be applied in different contexts.[34] Experts in printing, paper and typography were hired to visit the various IBM offices around the world and to lecture the staff on best practice.

Other companies did their best to make their own visual identities as distinctive and consistent as IBM's, but few succeeded. Nonetheless, the most memorable ones have become part of modern life. The 'golden arches' in the 'M' of McDonald's refer to the oddly shaped yellow arches that its co-founder Richard McDonald added to the architectural scheme for the fast-food chain's first drive-in restaurant in Phoenix, Arizona, in 1953. Having judged the original proposal to be dull, he decided to liven it up.[35] Then there is the secret joke in FedEx's logo. Next time you see that symbol, look at the shape between the letters 'E' and 'x'. It is an arrow, which seems entirely apt for a business whose purpose is to move things quickly from one place to another.[36] There is even an optically illusory effect when the logo moves. Watch a FedEx truck as it pulls away, and note how much faster it seems to go when you focus on the 'hidden' arrow. The same thing appears to happen when you see a Nike 'swoosh' in motion on the side of a pair of running shoes.[37]

The swoosh, like McDonald's golden arches and FedEx's secret arrow, is instantly recognizable, largely because we see it so often, always looking more or less the same. But corporate thinking no longer favours homogeneous logos to the degree that it did when Rand was imposing his visual will on IBM and wrangling with Steve Jobs over the position of the '.' in 'Steven P. Jobs' on his NeXT business card.[38] A new school of thought argues that such symbols may be more effective if they can be adapted to suit different purposes and contexts. There have always been examples of these dynamic identities, but they were relatively rare until recently.

Probably the most famous is the Michelin Man, alias 'Monsieur Bibendum', who was designed in 1898 by the French illustrator Marius Rossillon, known as O'Galop, as an emblem for the tyre company owned by the brothers Édouard and André Michelin. Several years before, they had spotted an oddly shaped pile of tyres during a visit to the Lyon Universal Exhibition, and Édouard had remarked to André that it looked like a man.[39] The brothers remembered this when they were looking at O'Galop's work and spotted a poster for a German brewery with a portly Bavarian drinking beer from a tankard engraved with the words 'Nunc est Bibendum'.[40] The boozy Bavarian bore a distinct resemblance to the humanoid pile of tyres the Michelins had seen in Lyon, and they asked O'Galop to draw a man made from tyres as a corporate motif for Michelin, then christened the result 'Monsieur Bibendum'.

At first the Michelin Man appeared on posters, then the

brothers hired actors to dress up as him for special events. Wherever he appeared, Monsieur Bibendum donned a sybaritic guise: often smoking a cigar, or sipping champagne. Soon he was depicted cycling through the French countryside, driving to a picnic, ambling off for a round of golf and bound for an Alpine ski trip. Most of his adventures involved travel, always in a contraption sporting Michelin tyres, and he exuded the bonhomie of a good sport who relished life's pleasures. As the company expanded into other countries and introduced new products, such as maps, travel and restaurant guides (all of which were intended to encourage people to drive, and thereby to buy more tyres), Monsieur Bibendum rose to the occasion by donning suitable attire: a Stetson for a visit to the United States, and a fez for a trip to Turkey.

But Michelin's mutating identity remained an exception, until the video music channel MTV went on air on 1 August 1981 with a corporate logo which was designed to seem different every time – or almost every time – the audience saw it, just as they would expect the channel to be playing different music whenever they tuned in. It was the work of three young designers who worked together as Manhattan Design in a tiny room on the corner of Sixth Avenue and Eighth Street in New York. One of the three, Frank Olinsky, had a childhood friend who worked for MTV's owner, Warner Amex, and invited them to submit a design proposal for the new channel's identity.

They decided early on that the core of the logo would consist of a large 'M' accompanied by a smaller 'T' and 'V', and that the effect should look spontaneous, rather than corporate. When Olinsky was a boy, one of his favourite TV shows was the children's cartoon series *Winky Dink*, which made an early stab at interactive television by asking viewers to intervene at key moments by drawing on a sheet of plastic film placed over the television screen to suggest how the star of the show, an affable character called Winky Dink, might get out of trouble. He decided to do something similar with MTV's logo. After blowing up a photocopy of the 'M', he took it out into the stairwell and added the 'T' and 'V' with a can of black spray paint so that it looked like the graffiti tagged on the New York streets. His friend took the result to show his bosses, and relayed their comments to Manhattan Design. Some of their suggestions were implemented, such as adding the words 'Music Television' beneath the initials; but others were rebuffed, including an insistence that any variations of the identity should be limited to an 'approved' palette of colours. Seibert and the designers argued that MTV should give carte blanche to all of the designers, artists, animators, film-makers and illustrators who would customize the logo over the years by depicting it covered in fur, dripping with paint, bursting into flames, splattered with blood or frozen in ice.[41]

The logo's success showed how appealing a dynamic identity could be, especially for MTV's young viewers, who had grown up in a visually saturated culture that gave them shorter attention spans than their parents, and heightened visual expectations. The work of cultural theorists like Roland Barthes and Jean Baudrillard had created an intellectual blueprint for deconstructing imagery in the 1960s and 1970s, but by the 1980s even people who had never heard of them or their writing could work it out for themselves, often instinctively. Not only had they become adept at decoding the messages encrypted in visual imagery (just think of how much more discerning we have become about typography since we have been able to choose typefaces from the Fonts menus of our computers), but their response to signs and symbols was very different. Rather than regarding homogeneous images as reassuringly authoritative and familiar, they dismissed them as dull; and far from being flustered or irritated by mutating imagery, they found it exhilarating.

MTV made its debut a decade before the World Wide Web went live in 1991. Once the Internet era began, attention spans became even shorter, visual awareness higher and the desire for distraction greater, all of which made dynamic identities more appealing. Digital technology also rendered them cheaper and easier to produce. It is easy to underestimate how brave Michelin was to plump for the constantly morphing Monsieur Bibendum as its motif in the late 1800s. In the era when corporate symbols were generally seen in print – whether on posters, company stationery, newspaper and magazine ads, or on signage – they were extremely expensive to make, especially if there was more than one version. Quality control was a problem too. The more variations there were, the greater the difficulty of ensuring that they were printed and installed to the same standard. But if corporate symbols are intended to be seen primarily online, like Google's and Twitter's, they can be adapted quickly and easily, at very little expense.

For online brands like those, this is just as well, because they have few other opportunities to engage with the people who use their services. No helpful sales assistants or attentive waiters. No friendly postmen, couriers or receptionists. For most of their customers, subscribers or visitors, the only hints as to what those organizations are like, and what they stand for, are whatever they see on their computer or phone screens. It stands to reason that the impression is likelier to be more favourable if they are intrigued or entertained by what they find there, rather than looking at the same thing repeatedly. What would you rather see? An identical set of letters spelling the name Google, or the naffest of its holiday logos? I would prefer to wince at the latter.

Saks Fifth Avenue, the City of Melbourne in Australia, Aol., the Netherlands Architecture Institute in Rotterdam, the Brooklyn Museum, the Casa da Música in the Portuguese city of Oporto, Channel 4 Television, Haus der Kunst in Munich, the London 2012 Olympic Games: for better or worse (in the case of the London Olympics), more and more organizations have plumped for dynamic identities. Even companies that still favour consistency have become a little more lenient. The jauntily illustrated penguin symbol that appears on Penguin's books has looked more or less the same over the years, with occasional variations, except for the Great Food series of paperbacks featuring food writing, which also sports a livelier version of the company motif, dancing a jig and defying anatomical logic by holding a knife with the tip of one wing and a fork with the other.[42] Not all organizations suit hybrid identities, notably the ones that we prefer to think of as being trustworthy, dependable and unchanging. As the Canadian designer Bruce Mau put it: 'MTV has a dynamic identity because they are dynamic, and I want them to be. But I don't want my bank to be dynamic. I want them to be conservative and radically stable.'[43]

Other dynamic identities have faltered for practical reasons, such as British Airways' ill-starred attempt to depict itself as an international, rather than a national, brand in 1997 by commissioning artists from all over the world to create ethnic emblems for its tickets and tail fins. One problem was the public outcry in Britain, where the new strategy proved unpopular. The former prime minister Margaret Thatcher disliked it so much that she draped a handkerchief over a model of a BA aircraft at an official function. Another difficulty was that air traffic controllers reportedly found it difficult to identify BA planes, because they all looked different. BA's arch-rival Virgin Atlantic took advantage of its plight by introducing BA's old symbol, the patriotic Union Jack, to its livery, and BA eventually dropped its controversial identity.[44]

Even the most successful hybrid identities find it hard to overcome the challenge of finding a distinctive name and descriptive symbol or set of symbols in an age when there are already so many. The surfeit of identities explains the plethora of made-up, sometimes silly names. Why call an insurance company Aviva? A management consultancy Accenture? A camera the Exilim? A car the i-MiEV? Another car the Th!nk? Not that the only foolish names are made-up ones. An irredeemably bland apartment block being built near my home has the howlingly inappropriate name Avant Garde Tower.[45] A hotel in the Bloomsbury area of London once sought to cash in on local literary lore by dubbing a restaurant 'Virginia Woolf's Burger Bar' in a nod to the novelist, who had lived nearby, and later rechristened it 'VW's Brasserie'. British design buffs still wince at the memory of the naming of the first large hovercrafts to go into service between Britain and France in the 1960s and 1970s. The French vessel was christened *Jean Bertin* in honour of a pioneering hovercraft design engineer, and the British one *Princess Margaret* after the Queen's younger sister, who was not noted for her interest in either design or engineering. But made-up names often sound even sillier than 'borrowed' ones. After all, if you have been given the chance to invent a name from scratch, why choose something stupid?

Another legacy of the dearth of – and frenzied competition for – original names is the craze for adding punctuation marks and other irrelevant symbols to corporate names and jumbling upper- and lower-case letters together, seemingly illogically. It is true that having taught ourselves to converse in the new digital languages of tweeting, texting and emailing, we have become more permissive about grammar and spelling, and more adept at using symbols on our keypads and keyboards as abbreviations. But some of the results are risible. Vélib', the Parisian cycle hire scheme, is a victim of this trend, as are the Th!nk and another electric car the G-Wiz. The i-MiEV suffers on two fronts: so do Aol. and Toys 'R' Us.

Yet it is still possible to design new symbols that are intelligent and inspiring, as the anti-capitalist activists in the Occupy movement proved in autumn 2011.[46] If anyone had written a design brief for Occupy's visual identity, it would have sounded impossibly daunting: to create a series of signifiers which would be distinctive and memorable, yet versatile enough to reflect the geographical and political diversity of a leaderless movement composed of hundreds of disparate groups embracing different causes. Whatever was chosen needed to work on home-made banners in Occupy's camps and protest marches, as well as on the Internet, particularly on Twitter, Facebook and the other social media sites with which local groups rallied support. The clinchers were that you would not be able to force any of those groups to adopt whatever you came up with, because they were free to choose whether to do so, and you would have to conceive and execute the identity for a pittance, ideally for less.

The original group picked a name, Occupy Wall Street, which could be adapted to suit any location: Occupy Winnipeg, Occupy Warsaw, Occupy Wellington, and so on. All of the subsequent groups took that name, but were free to pick and choose which other elements of Occupy Wall Street's identity they wished to adopt. Among the most popular choices were the slogans, such as 'We are the 99%'. Originally a reference to the concentration of personal wealth in the United States among the richest one per cent of the population, but versatile enough to apply to other countries, it also explained a complex economic concept clearly and persuasively, but was concise enough to be included in tweets without breaching Twitter's one-hundred-and-forty-character limit. Similarly, lots of Occupy groups took up variations on the slogan 'Sorry for the inconvenience. We are trying to change the world.' The wording may have varied slightly, but the meaning was consistent, as was its wit.

When it came to choosing visual symbols, the most ubiquitous ones included a traditional motif, the raised fist, that rooted Occupy within the trajectory of past protest movements, and the #, which seemed resoundingly contemporary, thanks to its reinvention as a Twitter identification tag. Not only did Occupy emerge with an original, versatile and stunningly effective identity, it treated the # to yet another new role, this time as both a descriptive and an instructive motif, which was equally effective and honest in either role.

An Objet Thérapeutique, designed by Mathieu Lehanneur, to help people to take the correct doses of prescription drugs

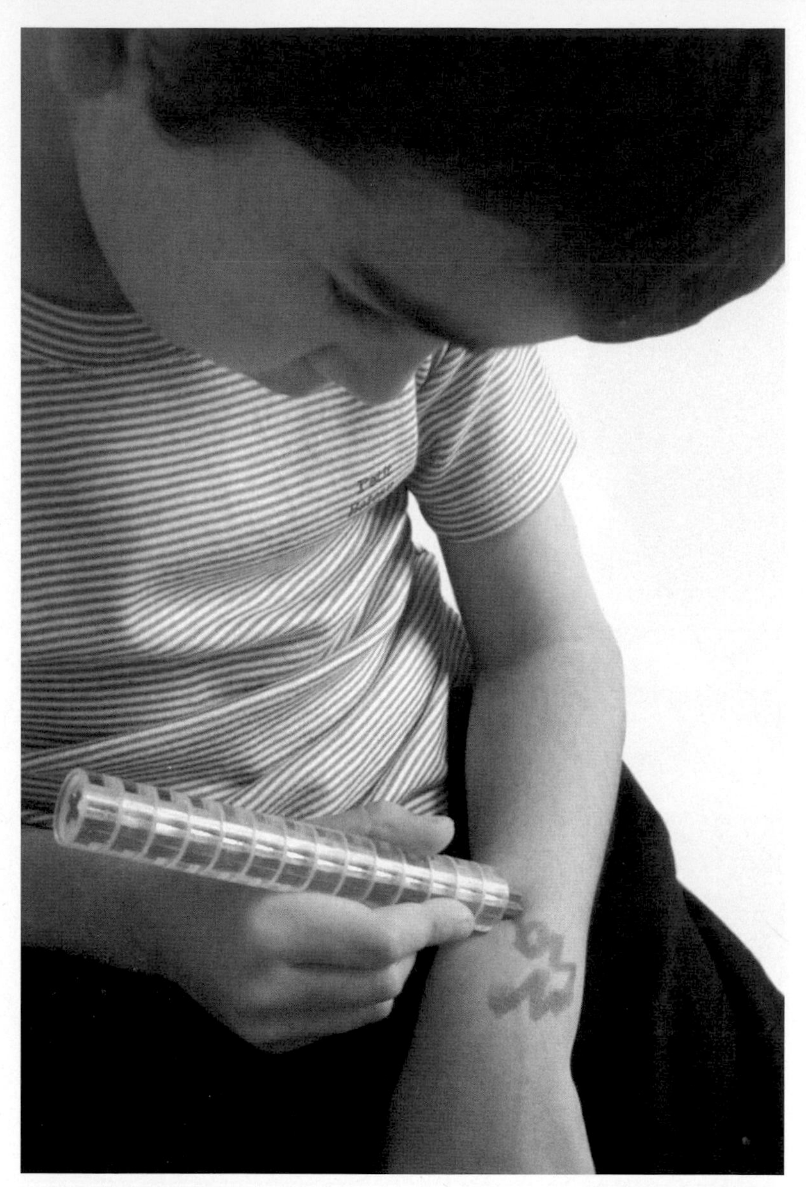

An Objet Thérapeutique
pen that dispenses
analgesic for acute pain

Each dose in a course of
antibiotics is peeled off this
Objet Thérapeutique in a layer

Mathieu Lehanneur's 'Tomorrow Is Another Day' in the palliative care unit of the Diaconesses Croix Saint-Simon Hospital in Paris

9 When a picture says more than words

Information is only useful when it can be understood.
— Muriel Cooper[1]

Charles Booth was a fortunate man. Born into a prosperous family in Liverpool, he inherited a substantial sum in his early twenties and invested it in setting up a new shipping line between England and Brazil. As Anglo-Brazilian trade flourished, so did his venture, making Booth even richer. Able though he was at business, his passion was politics, and during the 1865 general election he volunteered to canvass for the Liberal Party candidate in Toxteth, one of the poorest parts of Liverpool. Knocking on door after door to speak to the residents, Booth was horrified by the desperate conditions he found there, and flung himself into a campaign against poverty.

Having begun his relief work in Liverpool, Booth continued it in London, after moving there in 1875 following his marriage to Mary Macaulay. The squalor of London's frenziedly expanding slums, where crime, disease and hardship were rife, was a national scandal at the time. By the 1880s, Booth's fellow anti-poverty campaigners were claiming that one in four Londoners was living in intolerable conditions, but based on his own knowledge of the slums, he suspected this figure was exaggerated. Concerned that even an accidental inaccuracy could inflame public opinion at a sensitive time, Booth decided to ascertain the true extent of hardship in the city.[2]

Helped by Mary, he assembled a team of volunteers, mostly recent graduates, including the political economist Clara Collet and the sociologist Beatrice Webb, to assess the income and social class of the residents of every street in London, starting in the poorest neighbourhood, the East End.[3] Having spoken to the local people, Booth and his researchers consulted policemen, councillors, clerics and school boards to verify their impressions and the information they were given. For weeks on end, he left his home in plush Kensington to live in meagre lodgings in East London. Clara Collet did the same, renting a room in Whitechapel where she befriended the local women, many of whom made ends meet by working in factories during the day and as prostitutes at night, despite their fear of Jack the Ripper, who was terrorizing the area at the time.[4]

After analysing the data, Booth's team determined the socio-economic status of each street and recorded their findings in a series of maps published between 1886 and 1903.[5] The streets were depicted in different colours to signify the circumstances of their residents. A black street was occupied by the 'lowest class, vicious and semi-criminal', who lived 'the life of savages with vicissitudes of extreme hardship'. The people living on dark-blue streets were 'very poor' and in 'chronic want', while light-blue streets were for those who were 'poor' but not destitute. Purple signified slightly less poor, pink 'fairly comfortable' and red 'well-to-do'. Best of all was to live on a yellow street among the richest, grandest Londoners, as Booth did in Kensington.[6]

The first map was published as the 'Descriptive Map of East End Poverty' in 1889. A second set of maps covering all of central London came out in 1891, followed by a third series charting a wider area. Far from proving that the estimate of one in four Londoners living in poverty was an exaggeration, Booth's investigation revealed that it had underestimated the problem, and the correct figure was one in three.[7]

The most important component of Booth's poverty project was the rigour of its research, which may well have influenced public policy in whatever form it was presented. But the design decision to colour-code each street ensured that the information relayed by the maps could be readily understood by a far wider audience than the relatively small number of people who would have ploughed through an academic tome. Anyone could identify the pockets of deepest poverty at a glance, simply by looking for strips of black and dark blue; while the other colours acted as accurate and objective guides to the nuances of Britain's arcane class system. As a result, Booth's findings had greater impact on public opinion than they otherwise would have done, and stepped up the pressure on the government to clear the slums and replace them with decent housing. It is for this reason that his poverty maps are now hailed as exemplars of information design.

Translating huge quantities of complex, often contradictory information into a form in which it can be understood easily and accurately has been an important role of design from ancient Egyptian hieroglyphics onwards. Yet like other instinctive design exercises, including Ishtar's raised fist and the pirates' skull-and-crossbones, such feats were seldom recognized for their design merits.

The American statistician Edward Tufte has published a series of books hailing unsung examples of information design, starting with eleventh-century Chinese maps[8] and early seventeenth-century depictions of the solar system, including Johann Bayer's sky maps and the beautiful copper engravings in Galileo's book *The Starry Messenger*.[9] By the late eighteenth century, such visualizations were used increasingly as analytical tools. Among the pioneers were the Swiss-German scientist and mathematician J. H. Lambert, and William Playfair, a Scottish political economist who devised ingenious visual interpretations of statistics, including the first known bar chart, in his 1786 book *The Commercial and Political Atlas*.[10] Their work paved the way for nineteenth-century social reformers, including Charles Booth, to deploy visualization for political ends. The anti-slavery lobby used John Hawksworth's 1823 engraving of the layout of *The Vigilante*, a French slave ship captured by the British navy off the coast of Africa the previous year, to expose the heinous conditions in which African slaves were transported to North America. Nearly three hundred and fifty slaves were crammed into the hold of the ship in shackles, while the spacious captain's cabin occupied roughly the same amount of room as several dozen seated women. By depicting the ship in the dispassionate style of an architectural drawing, Hawksworth demonstrated the merciless brutality of the slave trade.[11]

Equally effective was a dot map devised by a London doctor, John Snow, to identify the source of a sudden cholera outbreak in the Soho area in September 1854. He produced a map of the affected area, marking the water pumps with crosses and the home of each cholera victim with a dot. As nearly all the dots were clustered around the Broad Street water pump, Dr Snow concluded that it was the likeliest source of the disease. His suspicions were confirmed when the handle was sent away to be tested and found to be contaminated. After its removal, the spread of cholera ceased.[12] As with Booth's poverty maps, Dr Snow's medical knowledge and the efficiency with which he analysed the data were the decisive factors in defusing the crisis, but it would have been much more difficult for him to have secured the institutional support he needed to tackle the problem had the information been presented less persuasively.

Tufte's own favourite example of historical information

design is a series of *cartes figuratives*, or flow maps, of famous military campaigns produced by a retired French engineer, Charles Joseph Minard, in the mid 1800s. One map traces the arduous journey of the army of soldiers and elephants led by the Carthaginian commander Hannibal over the Pyrenees and the Alps from Spain to northern Italy in 218 BC. By drawing Hannibal's army as a pale-brown band against a black and white geographical map of the mountains, Minard showed how many men he had lost, together with almost all of the elephants. Hannibal's army shrank from ninety-six thousand to twenty-six thousand men during the course of the campaign, and the band grows progressively thinner as the death toll rose.[13]

Minard's masterpiece was his map of Napoleon's ill-fated invasion of Russia in 1812. Using the same formula, it illustrates how France's Grand Army crossed the border between Poland and Russia in June that year with four hundred and twenty-two thousand troops. The pale brown band then dwindles to depict the tragic consequences of their four-month march to Moscow, and the one hundred thousand soldiers who survived it. A black band illustrates the retreat, including a disastrous crossing of the Berezina River in November in which hundreds more men died. Minard also traced the impact of the plunging temperature as the icy Russian winter set in. By the time the remaining troops crossed the border back into Poland in December, only ten thousand were left in a skinny black line, which is barely visible. Having personally witnessed the brutality of war, Minard hoped that his maps might deter future warmongers.[14]

As the logistics of daily life became more complicated in the early twentieth century, information designers stepped up their efforts to demystify them. One of the most ambitious initiatives was the International System of Typographic Education, or Isotype for short, an extensive range of pictorial symbols introduced in the late 1920s by the Austrian political scientist Otto Neurath in the hope of explaining important social, economic, political and scientific issues to people who had difficulty reading and writing.

Born into a wealthy Viennese family, Neurath had been radicalized in his early twenties when conducting academic research in poor rural areas of the Austro-Hungarian empire. He had no design training, but as a boy he was fascinated by the maps, charts and drawings in the books in his father's library, and by the Egyptian hieroglyphics he saw in museums.[15] Convinced that pictures could explain even the most complex ideas more effectively than words, he assembled a team of researchers and analysts, including his future wife, the mathematician Marie Reidemeister, to identify which symbols would be required to produce a comprehensive visual language and how they could be combined to express different issues. He then commissioned artists to draw them.

The most prolific of the artists was the German draughtsman Gerd Arntz. Like Neurath, he came from a privileged background but had become politically active, in his case after working in his father's factory. During the mid 1920s, he depicted the social and political problems of the era in a series of woodcuts. When Neurath saw them, he summoned Arntz to a meeting in Vienna in 1926, greeting him with a brusque: 'How much do you cost?' Arntz was to draw hundreds of symbols for Isotype. Each one was intended to depict an aspect of daily life in as simple a style as possible, with few lines and no perspective, in a limited range of colours. A lawyer was represented by scales of justice; a student by a man or woman writing at a desk; a farmer by a scythe; and a striking worker by a clenched fist.[16] Different symbols could be combined to convey more complex messages, and Neurath insisted from the start that they be used in multiples to indicate increases in size, rather than being made individually bigger. Isotype flourished in the progressive climate of 'Red Vienna' during the 1920s when the city was controlled by social democrats, who championed a stream of social and economic experiments. By the end of the decade, some twenty-five people were working on the project, and the Isotype team had close links with the Bauhaus, as well as with foreign artists and designers, such as El Lissitzky in Russia and Jan Tschichold in Germany.[17]

Neurath's researchers used the system to produce concise and accurate analyses of such complex phenomena as urbanization, economic development, the demographic differences between various countries and their relative readiness for war in the 1930s. The results were seen by millions of people in books, lectures and touring exhibitions during periods of intense political volatility before and during the Second World War. A series of wartime books on America and Britain used Isotype charts to analyse the distinctions between the two countries. In one book, entitled *Our Private Lives*, the charts reveal at a glance how highly educated the population of each country was, what type of meat they ate, whether they drank coffee or tea, how they spent their leisure time and which jobs they were likely to do.[18]

As Neurath liked to say: 'Words divide, pictures unite.'[19]

Other feats of early twentieth-century information design tended to be instructive rather than analytical. Among the finest examples was the London Underground Map, designed in 1931, initially on a speculative basis by a freelance draughtsman, Harry Beck.[20] Like many public transport systems, the London Underground had developed piecemeal, with new lines being built by different companies, and existing ones extended as new suburbs emerged. By the early 1930s, the companies had merged and the Underground sprawled across such a large area that it was increasingly difficult to squeeze all the lines and stations on to a geographical map, especially one which was legible enough for people to read and act upon when dashing through tunnels or across crowded platforms. Navigating the network was tricky enough for Londoners, including those with a reasonable knowledge of the city, but even more so for tourists.

Finding ways of improving the map was a constant concern for the Underground's management, and Beck took it upon himself to modify the existing version in his spare time. Having concluded that a geographical map was no longer feasible, he decided to adopt a diagrammatic approach, which would seem clearer and more logical to passengers, even if it risked distorting geographical reality. 'Looking at the old map of the Underground railways, it occurred to me that it might be possible to tidy it up by straightening the lines, experimenting with diagonals and evening out the distance between stations,' he said. To that end, he enlarged the crowded central area of the network and compressed the lengthy stretches of suburban tracks. He then connected all the stations on each line with straight lines, and positioned them as though they were similar distances apart, depending on whether they were in the city centre or the suburbs. Each line was allocated a different colour, and every station depicted as a circle: hollow rings for interchanges, and solid dots for the rest. Beck's 'tidying up' exercise even included straightening the course of the River Thames.

Having completed his first diagrammatic map, Beck showed it to his colleagues, who encouraged him to present it to the heads of the Publicity Department, only for them to reject it. Undaunted, Beck asked them to reconsider. A year later, he was summoned to see one of the managers, who said: 'You'd better sit down. I'm going to give you a shock. We're going to print it.'[21] Up until then, Beck had done all of the work on the map outside office hours and at his own expense. Only when the map was finally printed was he paid, and even then it was a nominal fee for producing the artwork.

The Underground placed an order for seven hundred and fifty thousand copies of the 'Diagram', as it was called internally, to be printed as a small, folded leaflet in January 1933.[22] The public response was positive, surprisingly so, given the radical nature of Beck's design. The distances between stations bore no relation to reality, nor did their positions. Angel station was shown on the same level as Old Street, rather than to the north. Victoria and St James Park were level too, although the former is south of the latter. Beck even modified the style of his chosen typeface in the interest of clarity, when it came to printing lengthy station names like Tottenham Court Road. Ever the stickler, he would have considered shortening them to be cheating. (Although he eventually admitted defeat with Ravenscourt Park, which had to be printed on three lines as Ravens- court Park on later editions of the map.[23])

Despite its calculated geographical gaffes, the 'Diagram' was instantly popular. The first batch of leaflets ran out, and a repeat order for one hundred thousand copies was placed after a few weeks.[24] The principal reason for its success was that it worked. The old Underground map really was confusing, and this one was not. Even first-time travellers could make sense of it. Beck's 'Diagram' was a rousing example of an ingenious design solution with instantly identifiable benefits for those who used it. No more late arrivals, missed meetings or stranding friends beside ticket booths because you had mistakenly taken the wrong line, missed a turning or ended up having to wait for a later train.

No wonder passengers liked it, especially as geographical inaccuracy must have seemed like a small price to pay for the map's efficiency. London is such a big city that few of its citizens were knowledgeable about its geography in the 1930s, nor are they today. Most of the people who used Beck's map would have been unable to guess the location of far-flung suburban stations like Rickmansworth, Hounslow, Wimbledon, Edgware, Richmond or Morden. If those stations were 'moved' to make the map clearer, passengers would not necessarily notice, so why should they object? (This was just as well, as Richmond in particular would be shuffled around repeatedly in subsequent revisions of the 'Diagram'.[25]) All the public knew was that the new Underground map seemed to make more sense than any supposedly accurate attempt to chart the geographical reality of London's labyrinth of medieval lanes, Georgian squares, Victorian terraces, modern suburbs and the twists and turns of the Thames.

The 'Diagram' was such a success that Beck was made a full-time employee of the London Underground, albeit not until 1937, five years after his design had been accepted. Ten years later, he resigned to take up a teaching post at what is now the London College of Communications, but continued to work for the Underground on a freelance basis until 1959. Throughout that time, he remained immersed in his map, constantly identifying improvements, such as changing the colour-coding in 1934 to make the Central Line more distinctive (it switched from orange to red),[26] and accommodating structural or operational changes to different parts of the network. Fiercely protective of the 'Diagram', he often found himself at odds with colleagues in his efforts to stave off what he saw as damaging interventions. Ken Garland's book *Mr Beck's Underground Map* recounts Beck's tireless campaign to safeguard his work through verbal complaints, letters of protest and legal threats.

The crunch came in 1960 when, unbeknownst to Beck, the Underground commissioned a new map to include the soon-to-be-completed Victoria Line, the only entirely new line to have been built since the 'Diagram' had been introduced. The first Beck knew of it was when the new map was published, complete with abbreviated station names – Trafalgar Sq., Liverpool St., Bow Rd. – and other graphic howlers such as splitting the name 'Aldgate' by placing 'Ald' on one side of the Circle and Metropolitan Lines and 'gate' on the other. (Unsurprisingly, it failed to solve the Ravens- court Park problem.) Up until then, Beck had been expecting to revise his own 'Diagram' to accommodate the Victoria Line, and had been looking forward to doing so. He was furious.[27]

Beck began a long, ultimately fruitless battle to try to persuade his old bosses to allow him to redesign the map. He lost the campaign and never worked for the Underground again, yet the spirit, at least, of his original 'Diagram' has survived, as one of the most popular emblems of London and a stunningly efficient way of navigating an otherwise baffling city. Other cities have adopted diagrammatic subway maps, including Tokyo, Milan and Amsterdam. But Beck's approach has proved most effective in places that are structurally similar to London: agglomerations of villages that have merged over the centuries to stretch across vast geographical areas, with densely clustered towers surrounded by low-lying suburbs. Tokyo is like that, but Paris is not. The Paris Métro invited Beck to design a diagrammatic map, only to realize that it would not be suitable for a smaller city with a more orderly, radial layout. The idea was dropped quietly.[28]

The outcome of the New York subway's flirtation with a Beck-inspired map turned out to be much noisier. In 1972, it introduced a diagrammatic map devised by the Italian-born designer Massimo Vignelli, who had advised the New York Transit Authority on design since the late 1960s.[29] He relished the challenge of 'tidying up' the tangled subway system into a seemingly rational network depicted in twenty-two colours, one for each line, with every station shown as a dot. Design purists still reminisce fondly about Vignelli's design, but a vociferous group of New Yorkers loathed it from the start, and voiced their complaints loudly.[30]

The problem was that New Yorkers were more familiar with the geography of their city than Londoners were with theirs; at least, they thought they were. Very few subway passengers were familiar with the length and breadth of the network, including all the suburbs. But the central area of New York is small enough to be walkable, unlike London's, and the 'grid' system of numbered avenues and cross-streets introduced to Manhattan by the Commissioners Plan of 1811 is so orderly that New Yorkers are likelier to think that they know their way around, even if they do not.[31] They also have the comfort of knowing that, thanks to the grid, if the worst came to the worst, they could work out where to go. Their confidence significantly reduced the perceived benefits of a diagrammatic subway map, and made New Yorkers less inclined to forgive its geographical anomalies. Even those that managed to do so found it hard to ignore the most obvious ones, especially the glitch that Vignelli's critics claimed was the most egregious of all, depicting Central Park as a square, as opposed to its actual shape of an elongated oblong, at least three times bigger than the diagrammatic version.

Eventually, the transit authority buckled and Vignelli's map was abandoned in 1979. He continued to defend it, arguing that, rather than going too far in terms of ignoring geographical reality, his mistake was not to have gone far enough. Vignelli had made a few geographical references on the map: to the locations of Manhattan, Brooklyn and the Bronx, for instance. In retrospect, he felt that his chances of success would have been higher had he ignored them. When interviewed by Gary Hustwit for his 2007 documentary film *Helvetica*, Vignelli explained ruefully that the world was divided between 'visual' people, who had no difficulty reading maps, and 'verbal' ones, who often did. 'Verbal people have one great advantage over the visual people,' he groaned, 'they can be heard.'[32]

Quite so, though if there is a moral subtext to the saga of those subway maps, it is that the rest of us are happy to allow designers to simplify life's complexities, even if it involves straying from the truth, providing that we enjoy the result. New Yorkers rejected Vignelli's map because they found his attempt to edit out their city's idiosyncrasies to be dull and soulless, whereas Londoners still like the look of the neat, navigable city depicted by Harry Beck. Indeed, most of them are so familiar with the latter-day version of his 'Diagram', and so confident that they can trust it, that they do not think of it as being inaccurate. When the London Eye opened beside the Thames in 2000, many Londoners saw a 'bird's eye view' of their city for the first time, and expressed astonishment that it looked so different from the Underground map. The river turned out to be wigglier than they had thought, and many people had no idea that some of the landmarks on the South Bank, including Tate Modern and Waterloo Station, were actually north of the Houses of Parliament and Westminster Abbey, both of which are on the northern bank of the Thames.

It takes considerable skill to pull off such a convincing act of (entirely well-intentioned) design deception, but not even a designer with Harry Beck's flair for condensing a blizzard of information into brevity and clarity could have risen to the challenges facing information designers in the digital age, at least not if he had stuck to print. Millions of people still use printed maps, charts, pictograms and diagrams like those devised by Beck, Neurath, Minard, Snow, Hawksworth and Booth to help us to make sense of the world, but the unprecedented speed, scale and complexity of modern life means that we need new forms of information design too.

The central issue is the colossal increase in the volume of data that computers produce and store, the 'data deluge', as the *Economist* has called it.[33] 'Moore's law' is to blame. In 1965, Gordon E. Moore, a founder of Intel, the US chip manufacturer, calculated that the number of tiny transistors which could be packed on to a silicon monocrystal microchip had doubled every year. He also predicted that this would continue for at least a decade. Dubbed 'Moore's law', his forecast has remained accurate ever since, and as microchips have become more powerful, so have the machines they operate.[34] As a result, the volume of traffic on the Internet increased eightfold in the five years to 2011, and is expected to rise threefold between then and 2016, when it will reach over a zettabyte, such a huge quantity of information that the word was only invented in 1991.[35] In practical terms, this means that the amount of data generated by the Internet is growing considerably faster than the network's ability to carry it.

Not only has computer processing power escalated, but software has become more efficient. The most sophisticated systems now take days to complete projects which would have taken years until relatively recently, or proved impossible to execute. When the Sloan Digital Sky Survey's giant telescope in New Mexico started operation in 2000, it amassed more data in its first few weeks than in the entire history of astronomy up until that point. And it produced as much information in the whole of its first decade as the new Large Synoptic Survey Telescope now being built in northern Chile will be able to amass in just five days. Similarly, it is now possible to decode the human genome within a week, compared to the ten years of computer analysis required when it was first completed in 2003.[36]

These advances are aggravating data deluge, but may also offer a solution, if the same increases in processing power and software efficiency are applied to developing new ways of managing the maelstrom of information. A group of software designers has responded by developing computer programs which fulfil the old-fashioned function of information design by crunching through the data and reinterpreting it in a way that is legible and useful to the rest of us, without over-simplifying it by introducing errors or losing nuances. The result is a new genre of dynamic digital images known as data visualizations: computer-generated images and animations that can be programmed to change in real time to track the information they are relaying.

Their origins lie partly in the research into computational aesthetics begun by László Moholy-Nagy in late 1920s Berlin when he experimented with moving images on home-made contraptions like the Light Space Modulator, which stupefied customs officials when he and his family were moving around Europe in the 1930s.[37] His assistant at the time, György Kepes, continued this work in the United States, first when teaching in

Chicago with Moholy, and then in his own research at the Massachusetts Institute of Technology, where he spent most of the rest of his career.[38] Kepes's work there and his books, such as 1944's *Language of Vision*, had a profound influence on colleagues such as Nicholas Negroponte and Muriel Cooper, and through them on future MIT students, including the software designers John Maeda and Lisa Strausfeld.[39]

Not all the aesthetic pioneers of data visualization came from such a nurturing academic environment as MIT. Many of them were self-taught amateurs working in isolation. Computers had been restricted to military use until the end of the Second World War, when they were introduced to universities and research laboratories, where scientists and mathematicians started to experiment with them to create visual effects. Mostly they did so by attaching mechanical arms to guide pens or pencils across sheets of paper. Their goal was aesthetic, to find out what sort of images a machine would produce of its own volition. Throughout the 1950s and 1960s, artists gradually became aware of the unusually precise nature of the imagery generated by computers, as did people from other fields.[40]

Among them was a lecturer at the University of Manchester, Desmond Paul Henry.[41] The university was responsible for important early advances in computer science, not least in building the world's first stored-program computer, the Manchester Small-Scale Experimental Machine, which was nicknamed 'Baby' when it was completed in 1948. Henry took a keen interest in such developments, but as an observer, not a participant. He taught philosophy at the university, and had discovered computers as a technical clerk for the Royal Electrical and Mechanical Engineers during the Second World War. In 1951, while browsing around the second-hand bookstalls on Shudehill in Manchester, Henry went into an army surplus warehouse and spotted an old Sperry bombsight computer used by British bomber aircraft to calculate when to release their bombs. He bought it for £50, a substantial sum at the time, and rigged it up to guide ballpoint pens and, later, technical tube pens across paper. After his first machine died, Henry recycled some of the components into a new one and subsequently repeated the process to build a third. He exhibited his mechanical drawings at local art galleries in Manchester and also in 'Cybernetic Serendipity', the 1968 survey of computer art at the Institute of Contemporary Arts in London, which later toured to the United States. When his computer returned from the tour, it was broken, and although Henry built two more machines, his interest in computer art faded and he concentrated on philosophy, becoming an authority on medieval logic.[42]

Touchingly naïve though they look now, the computerized images produced by Henry and fellow pioneers, such as Charles Csuri, Harold Cohen, Roman Verostko and Vera Molna, must have seemed extraordinary at the time. The lines, forms and repetitions are too improbably precise to have been made by hand or even by conventional machines. Kepes's and Cooper's experiments at MIT had a similarly gauche quality, yet all of these early works have helped to produce the intensely sophisticated digital imagery we see today.

Data visualization is also rooted in scientific research from the 1980s onwards. Science was one of the first fields to be affected by data deluge, because computers had pumped out ever-increasing quantities of scientific data since the 1960s. This gave scientists the information required to grapple with new and more difficult problems, but there was a growing realization that they needed to find more efficient ways of testing their assumptions and of explaining the results of their research to non-scientists, in particular politicians, whose support would be needed to tackle environmental issues. By the mid 1980s, a growing group of scientists was interested in exploring the possibility of using computer graphics to translate their theories into images.

The National Science Foundation in the United States convened a Panel on Graphics, Image Processing and Work-stations to discuss the issue in 1986.[43] The following year, the panel published a report on 'Visualization in Scientific Computing' in which it outlined the potential of computer graphics to aid scientific research and to disseminate it more widely.[44] The NSF allocated funds for suitable facilities to be installed at the super-computing centres which had recently opened throughout the United States. One of the first successes was the Vapors project, which demonstrated how damaging smog was threatening to become in southern California. The computer-generated image of how badly the region would be affected was so realistic that the authorities agreed to introduce more stringent pollution controls.

The technology developed to produce scientific visualizations was then applied to other fields including engineering, architecture and product design, where it was used to produce accurate digit-al models of finished projects. More recently, it has been deployed as an information design tool in the form of data visualizations. Some of the earliest examples dealt with developments in the digital domain, such as mapping the flow of traffic across the Internet, which would be impossible to represent accurately using traditional printed media.[45]

An important breakthrough was the development of Processing, a computer programming language initiated by the American software designers Ben Fry and Casey Reas in 2001, when they were colleagues at MIT Media Lab, where Muriel Cooper's protégé John Maeda was among their teachers. Processing was designed specifically to enable technologists to express themselves visually, while being simple enough for designers, artists and architects with no experience of programming to use it to produce images, models and animations.[46]

Fry harnessed Processing to produce a remarkably clear and succinct explanation of the genetic difference between human beings and chimpanzees in 2005, shortly after the publication of the first analytical comparison between human and chimp genomes. The differences are very slight, and a key distinction is believed to be the FOXP2 gene, which is linked to language. Fry's visualization created a photographic-style image of a chimpanzee's head from seventy-five thousand digitally generated dots, each representing one of the letters in the gene, with just nine dots highlighted in red to identify the only significant differences between the coding of the chimp and human versions of that gene.[47] You could read several weighty academic tomes on the subject and still not grasp it as clearly as you can in that image.

Data visualizations have since been used to illustrate everything from the progress of natural disasters such as earthquakes, floods and tsunamis; the often circuitous process of garbage disposal; the location of emergency calls in particular cities; recurring themes in the campaign speeches of US presidential candidates; to the flow of asylum seekers across the globe. The new interactive maps produced by the New York subway no longer simply show passengers what the network looks like, as the traditional printed one does, but, in the case of 'The Weekender', a digital map posted on its website, also alert them to delays, closures and seasonal hitches, such as leaves on the line, as well as suggesting the most efficient way of getting from one station to another between Friday night and early on Monday morning.[48] 'The Weekender' was also a personal triumph for Vignelli, because MTA officials invited him to reinterpret his diagrammatic map for the project, more than thirty years after their predecessors had axed it.

Hundreds of websites have emerged, symposia convened and books published on data visualization. Ferocious debates have raged over whether too many visualizations are strong on style but weak on substance, because their designers are obsessed by aesthetics or technology to the detriment of other qualities. But enough visualizations have proven themselves to be useful, and some indispensable. Often they are the work of young designers and programmers working independently in grungy studios, but visualization swiftly became a standard tool for government departments and bastions of 'old media' such as *The New York Times*, the *Guardian* and the BBC.

Nor has design's co-option of data visualization detracted from scientific progress. On the contrary, it has fostered greater collaboration between scientists and designers to the benefit of both fields, while scientific developments in visualization have continued to advance in long-running programmes such as the Blue Brain Project, for which a team at the École Polytechnique Fédérale de Lausanne is building a virtual brain in a supercomputer, thereby enabling scientists to learn more about and possibly develop new ways of treating neurological disorders and to study the impact of new drugs.[49]

Visually luscious and logistically dazzling though data visualization can be, there is still a place for old-fashioned forms of information design. Some of the most interesting recent examples are the work of the Dutch designer Joost Grootens. Having originally trained as an architect, he then worked as a multimedia designer before deciding to apply his knowledge of digital media to developing books which would be useful and appealing in the online era. He began by producing a series of atlases for the Rotterdam-based publisher 010 in which he reinvented the traditional atlas by devising new types of maps and other visual means of explaining the implications of its content.

His greatest feat is a book on his own work, published in 2010 as *I swear I use no art at all: 10 years, 100 books, 18,788 pages of book design*. In it, Grootens describes the process of designing each of those hundred books and the evolution of his work, partly in words, but mostly visually, in maps, charts, grids, infographics and indices. There are floor plans of every office and design studio he has worked in with coded numbers to indicate who sat where. A map of northern Europe shows the cities where he has held meetings, printed books and staged book launches. He has included examples of every typeface he has used, photographs of the binding of each book and coded diagrams of the layouts. There are also lists of all the authors, publishers, printers and colleagues with whom he has collaborated, and a flow chart which shows how one project led to another through meetings and word of mouth in a stunningly concise account of the random cocktail of tactics and coincidence that determine most design careers.[50]

Convinced that people stop looking at books as soon as they start reading the words, Grootens presents the visual information first. He proves the point in *I swear . . .* by printing the text in reverse on the few reproductions of pages taken from the books he has designed over the years. And he forces his readers to search for the things they probably most expect to find in a book about book design – images of the covers – by 'hiding' them inside the folded pages of an essay, knowing that the hunt for those pictures will make them seem more precious than any number of words could have done.[51]

MAP DESCRIPTIVE OF LONDON POVERTY, 1898-9
(IN 12 SHEETS)

SHEET 5.
EAST CENTRAL DISTRICT

THE STREETS ARE COLOURED ACCORDING TO THE GENERAL CONDITION OF THE INHABITANTS, AS UNDER:—

Lowest class. Vicious, semi-criminal.
Very poor, casual. Chronic want.
Poor. 18s. to 21s. a week for a moderate family.
Mixed. Some comfortable, others poor.
Fairly comfortable. Good ordinary earnings.
Middle class. Well-to-do.
Upper-middle and Upper classes. Wealthy.

A combination of colours—as dark blue and black, or pink and red—indicates that the street contains a fair proportion of each of the classes represented by the respective colours.

East London in Booth Map 5 from Charles Booth's 'Map Descriptive of London Poverty, 1898–9'

An Isotype exhibition at the Gesellschafts und Wirtschaftsmuseum in Vienna, *c.*1929

Marie Reidemeister working on an Isotype chart in Vienna, *c.*1930

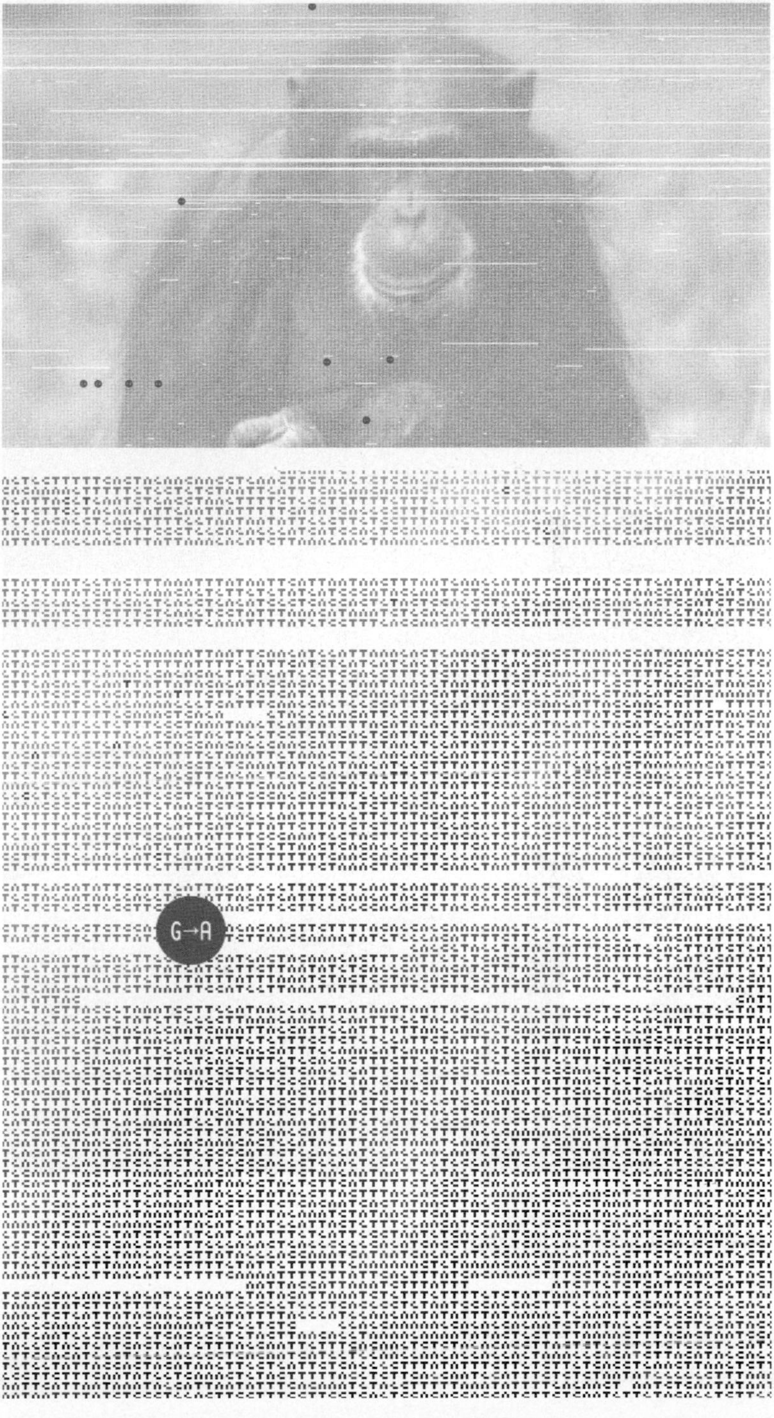

'Humans vs Chimps' data visualization designed by Ben Fry in 2005 using Processing software

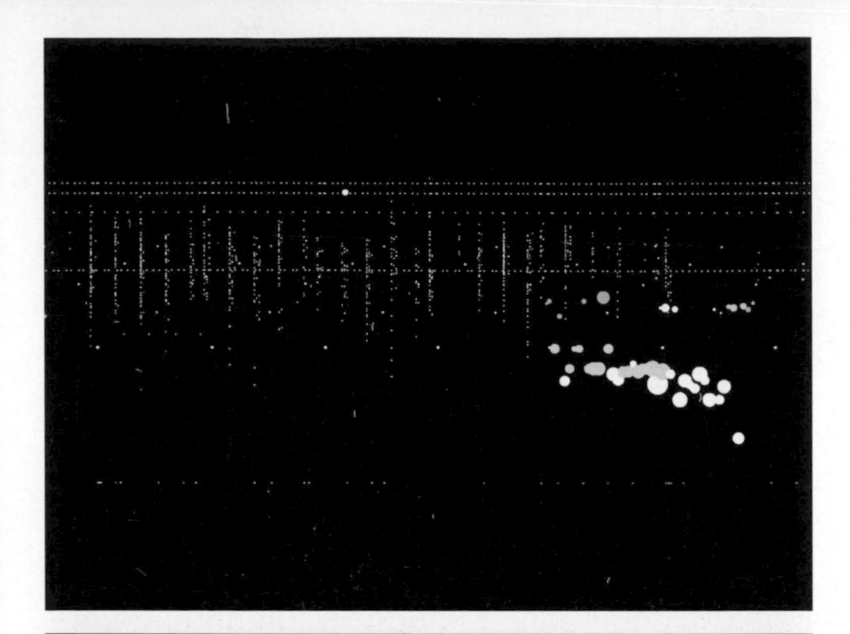

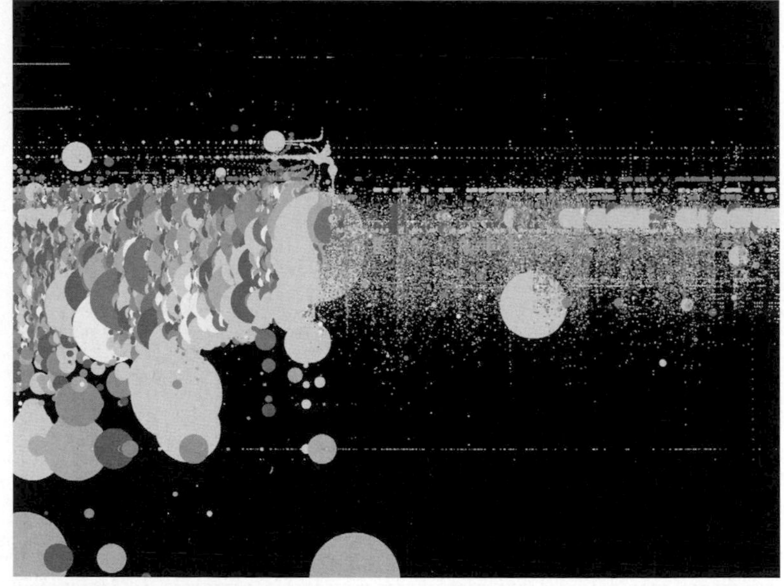

From Stamen's data visualization
of trading on the NASDAQ stock
market in New York on 24 July 2012

10 It's not that easy being green

Our great task is to bring man in scale again with the entire horizon of nature, so that he can sense it in all its wealth and promises, harmonies and mysteries. In ignorance and pride and by insecurity, we have severed ourselves from our broader background.
— György Kepes[1]

As the members of the Don't Make a Wave Committee drifted out of a meeting in Vancouver to plan an anti-nuclear protest voyage, one of them raised his fingers in a peace sign. 'Make it a green peace,' said another.[2] The phrase seemed so apt that they chose it as the name of the boat. An activist's son offered to design 'Green Peace' badges to be sold to raise money for the voyage, but as he could not squeeze both words on to them, he merged them into one – 'Greenpeace'.

Not only did the protest boat set sail for the US government's nuclear test site in the Aleutian Islands on 15 September 1971 as the *Greenpeace*, but the group decided to rechristen itself the Greenpeace Foundation.[3] By the late 1970s, 'green' had become the default name and symbolic colour for ecological activists all over the world. There were Die Grünen in Germany, Groen! in Belgium, Les Verts in France, and later De Grønne in Denmark, Federazione dei Verdi in Italy, The Greens in Australia, and dozens more. And why not? What could be more fitting as an emblem of environmentalism than green, the colour of nature in many countries, and of paradise in Islamic culture? Except that in the largely man-made world of design, green is often far from natural or paradisiacal.

The problem is that green is an elusive and unstable hue. From the Italian Renaissance to the Romantic movement, artists struggled to mix exact shades of green paint and to reproduce them accurately. Green dyes and pigments proved equally problematic in the industrial age, so much so that toxic substances were often used to stabilize them. Some of the green wallpapers made in the eighteenth and nineteenth centuries contained arsenic, prompting public outrage when people died from deadly fumes as they rotted.[4] Even today Pigment Green 7, a common shade of green in plastics and paper, includes chlorine. Another popular hue, Pigment Green 36, contains bromine as well as chlorine; and Pigment Green 50 has traces of cobalt, titanium and nickel. If potentially damaging pigments like those are used to dye plastic green or to produce green ink for printing on to paper, it will be impossible to recycle or compost those products safely lest they contaminate everything else.[5]

In other words, the colour green, the enduring symbol of ecological purity, is often not very 'green' at all. As Kermit the Frog sang on *The Muppets*, 'It's not that easy being green',[6] not least in design. The quest to build a safer, more sustainable society is an exhilarating opportunity for designers to make a positive impact on our lives. For many reasons, ranging from unalloyed altruism to glory seeking, they would be crazy not to grasp it. Yet their efforts to help us to live more responsibly have often been as troubled as the history of green itself.

There can be no doubt about the need to act. Even if they suspect global warming of being a fallacy cooked up by demented conspiracy theorists, no designer – or anyone else – can pretend to ignore the damage wrought by the degradation of the environment. All they need to do is look at a landfill site. It does not need to be one of the bigger ones, like the giant dump at Agbogbloshie in the Ghanaian capital, Accra, where discarded computers are shipped from Europe and North America to be burnt by scavengers, often boys, sent from their homes hundreds of miles away, in the hope of extracting scraps of copper, brass, aluminium or zinc from the debris.[7] Any landfill will do, because it is there that designers' work often goes to die in an ignominious death which can contaminate the land for decades to come.

Imagine landing a plum job with a stellar design team and working long hours, wrestling with the laws of physics to produce a gleaming new digital device, only to see desperate child workers foraging among its charred remains in a hellhole like the Agbogbloshie dump. Surely it would make you wish that you had worked differently. Perhaps by identifying more sophisticated materials and processes which may have made the product safer to dispose of. Or by trying harder to ensure that it could be recycled, rather than ending up somewhere as noxious as that Ghanaian tip, or on a latter-day version of Fresh Kills on Staten Island, New York, once the world's largest landfill site, two and a half times bigger than Central Park and taller than the Statue of Liberty.[8]

There are two ways of describing the outcome of that prestigious design job. One is that the designer helped to develop alluring and innovative products which people loved to use, and won critical acclaim as well as coveted prizes. Another is that he or she produced something which ended up failing to decompose in a bloated landfill site, or being dismembered in a fetid dump where child workers risked injury, disease and worse in the charred soil.[9] Both descriptions would be accurate, and could be applied to the outcome of many other design endeavours.

In Tim Brown's book *Change by Design* he describes how an IDEO design team developed a new children's toothbrush for Oral-B. It sold well, and the kids who used it seemed to like it. But one day the lead designer on the project was walking along a beautiful, deserted beach in Baja California, Mexico, when he spotted something colourful in the sand. It was one of the toothbrushes, which had washed up with the tide.[10] Hardly a devastating incident in environmental terms, at least not in isolation, but marine debris is a serious problem. One 'island' of floating junk known as the Great Pacific Garbage Patch or Trash Vortex has emerged in the Pacific Ocean and is twice the size of Texas.[11] No designer would like to think of his or her work ending up there, or as a blight on an otherwise unspoilt Mexican beach, and to consider what ecological damage it may have caused on its way.

Feeling outraged is the easy part. Designers generally agree on the principle that protecting the environment is important, and that they should try to help the rest of us to live responsibly, but they are likely to disagree about everything else relating to it. What does sustainability mean in design? What could – and should – it deliver? How can success be judged? Is it permissible to compromise, and if so, in which circumstances and to what degree? All these questions are fiercely disputed with different designers expressing contradictory views, often with equal conviction.

Consider the children's toothbrush. It was the result of a well-intentioned design exercise and would not generally be considered an environmentally hazardous product. What were the design team's responsibilities with regard to its ecological consequences? Some designers would argue that their duty would be discharged by producing a toothbrush which does its job so well, and is so robust and appealing, that a child will want to use it for as long as possible, rather than being tempted to throw it away or replace it. After all, they might say, how can a designer be expected to control what happens to a product once it has been sold, or the manufacturer or retailer, come to that? Surely, whoever ends up owning or using the toothbrush should take responsibility for it from then onwards.

Others would dismiss this argument as feeble, or downright irresponsible. Of course, designers must try to make things that last, but longevity, they contend, is not enough on its own. Designers must also strive to ensure that neither they, nor anyone else, need have any qualms about their work from the moment it was conceived until it is disposed of, and that its environmental impact will not simply be neutral, but positive.

What issues does this raise for the toothbrush? Is it made from recycled materials? Can it be recycled itself – safely and economically? How much water was used to manufacture it? And how much energy? Has the manufacturer or retailer provided a user-friendly way of sending the toothbrush for recycling? What about the packaging? Is it made from recycled and recyclable materials, with no toxic inks that might contaminate the groundwater system? Is it as compact as possible, to allow more toothbrushes to be packed into a box, and more boxes into a freight container, thereby minimizing the amount of energy required to ship them from place to place? And what sort of energy is that? Did the designer ensure that all these issues could be tackled cost-effectively to prevent bad practices from creeping back in?

If the ecological implications of designing a relatively straightforward product such as a toothbrush are so tortuous, how much worse will they be for something like a road transport system, which is embedded in labyrinthine networks of different industries, services, political and regulatory bodies, all with their own agendas and vested interests?

The case for redesigning road transport is indisputable, because the status quo is dangerous and dysfunctional. There are already some eight hundred and fifty million cars and trucks worldwide, nearly enough to circle the globe a hundred times if parked bumper to bumper. Most of them are powered by noisy, polluting internal combustion engines, which consume eighteen million barrels of oil and emit over seven million tons of carbon dioxide every day. The roads are so congested in many cities that average speeds are less than ten miles per hour. Over a million people worldwide are killed in road accidents each year, and many more are injured.[12] Yet the number of vehicles is still increasing: by two thousand a day in Beijing alone, as its citizens replace their bicycles with cars and trucks, at the same time as other cities are striving to foster the type of cycling culture that China is forsaking.

The first step is for the automotive industry to make a different type of car, by replacing mechanically controlled vehicles powered by internal combustion engines with electronically controlled ones fuelled by renewable energy. Technologically there is nothing to stop it from doing so, and the industry is moving in that direction, albeit slowly, and so far without having produced a car whose design is so compelling that it will transform the market as the Ford Model T did in the early 1900s, and the iPod has done for digital music and the Kindle for electronic books. Even if such a vehicle is finally developed, redesigning the car will not be enough on its own to make road transport safe and sustainable, because every other aspect of the system needs to be redesigned too.

As the number of energy-efficient cars increases, so will the demand for facilities where their batteries can be recharged or they can be refuelled with renewable forms of energy such as hydrogen. One option is to convert existing petrol stations for the purpose. Another is to adapt electronic street fixtures like phone kiosks, street lights and parking meters, as the Spanish government has done in Barcelona, Madrid and Seville.[13] It might also be possible to reduce congestion by introducing dynamic pricing systems for recharging and refuelling, and for parking tariffs too. Knowing that all those things will cost less when the roads are quietest could encourage drivers to use their vehicles then, rather than at busier times.

Then there is the design of the roads. Designating special lanes for bicycles and buses, or dedicated cycle superhighways, like those in Copenhagen, can help traffic to flow more efficiently.[14] Doing the same for energy-efficient cars may have a similar effect, while making those vehicles more attractive to motorists than wasteful gas-guzzlers. The flow of traffic might also benefit from replacing traffic lights with roundabouts, which often prove more effective at preventing accidents and unnecessary delays.[15] The design of any remaining traffic lights could be improved too. The most impressive ones I have seen were in the Chinese city of Tianjin. They consisted of rectangular strips of energy-efficient light-emitting diodes, or LEDs, which alternated between red, amber and green. At the start of each signal, the strip was fully illuminated. It then shrank in size, enabling drivers to estimate how long it would be before the signal changed, making them less likely to stress over possible delays, or to be panicked into braking or accelerating at the wrong moments. Finally, to help colour-blind motorists, many of whom see red, amber and green as similar shades of grey, the green signal shrank down towards the bottom of the panel to tell them to 'go', and the red shrank upwards for 'stop'.

The logistics of driving also need to be redesigned. In their book *Reinventing the Automobile*, William J. Mitchell, Christopher E. Borroni-Bird and Lawrence D. Burns proposed the development of the 'Mobility Internet', which would enable cars wirelessly to source and exchange information on driving conditions and anything else that might affect the journey.[16] The system could alert the vehicle and its driver to bad weather, traffic jams, accidents and sudden obstacles such as a fallen tree, or a chemical spillage from an overturned truck. The car would then be able to switch to a more efficient route before reaching the congested area and getting stuck there. The same technology could enable vehicles to anticipate and avoid collisions, as well as guiding cars along the road while the drivers were working, surfing the Web or resting. Such a system would be a boon to people with disabilities which might otherwise deter them from driving, and for frail elderly motorists. Scary though it sounds to allow a car to drive itself, we are already accustomed to riding in driverless trains and to flying in aircraft that land automatically. Driverless cars, like Google's, may well be safer by saving motorists and pedestrians from the danger posed by drivers who are drunk, and by those that fall asleep or allow their attention to wander.[17]

Design also has an important part to play in halting the growth of car ownership. Some of the most interesting exercises in design thinking have been the development of vehicle-sharing schemes, like Zipcar, and online services that enable people to check whether it will be possible to share a journey with another driver who is planning to take a similar route at the same time. Equally innovative are urban cycle-hire systems, such as Vélib' in Paris and Ecobici in Mexico City, which have had some success at encouraging people to switch to cleaner, healthier forms of transport.[18]

Curbing car ownership is a challenge all over the world, but is particularly urgent in rapidly expanding economies in Asia and Africa, where 'Smeed's law' has struck. Coined by the British statistician R. J. Smeed in 1949, Smeed's law suggests that whenever the number of new vehicles increases, so does the risk of accidents for their *ingénu* drivers. The newly built highways in Ghana are flanked by signs announcing the number of fatalities on those stretches of road and warning drivers to slow down. Yet there is still a powerful political attachment to car ownership as a symbol of economic virility in Ghana, and in other developing countries. A few years ago, I travelled along a recently completed Chinese motorway heading south of Beijing to Tianjin. It was lined by empty billboard hoardings, erected in the confident expectation that they would eventually be filled with advertisements. The motorway also sported the concrete skeletons of flyovers which were not yet connected to roads. They too had been constructed on the assumption that ring roads would be needed once Beijing and Tianjin sprawled outwards into new suburbs. By building the flyovers so far in advance, the Chinese authorities could avoid having to stop the motorway traffic during construction. Clearly, they expected the growth in car ownership to continue, and had invested accordingly, despite the rising toll of road deaths and pollution.

Think of all the people and organizations with some sort of financial, personal or political interest in all those issues. What a minefield to navigate in order to design a sustainable alternative to the dysfunctional status quo. And travelling by car is only one aspect of daily life which needs to be radically reconfigured if it is to be made sustainable. All the others will prove equally arduous, not least as the science of sustainability is so contentious, with acrimonious rows over many issues. I once attended a debate on marine biodiversity at a World Economic Forum conference, and arrived expecting to find the assembled scientists and environmentalists united in a common cause and determined to capitalize on its new-found prominence. Instead, they spent hours bickering about an academic squabble dating back to the early 1990s. No progress was made on that issue or any others that day. Intervening constructively in such combative territory is difficult for any discipline, including design. None of which means that designers should not try to tackle such challenges at a time when the environmental crisis is deepening and, on a positive note, consumers' expectations of sustainability are increasing and the practice of environmentally responsible design is gradually becoming less onerous.

One encouraging development is the cultural shift within the design community and in external perceptions of design. Traditionally, design was a dirty word to many environmentalists, who were prone to dismissing designers as the evil accomplices of ruthless capitalists in the despoliation of the ecosystem. Unfair though this was in some respects, it was not entirely untrue, not least as the commercial design industry proved so adept at concocting new ways of persuading people to buy more stuff, regardless of whether they needed it or were likely to want it for very long. Since the Industrial Revolution, the design profession has been steeped in consumerism and in believing that innovation is a force for good and that the new will invariably be better than the old. The memoirs of many twentieth-century designers are replete with references to design's power as a sales tool, but lack any acknowledgement of their social or environmental responsibilities. Raymond Loewy was not alone in believing that the goal of every industrial designer should be to keep 'his client in the black'.[19] Even the more thoughtful Henry Dreyfuss wrote: 'If people are made safer, more comfortable, more eager to purchase, more efficient – or just plain happier – by contact with the product then the designer has succeeded.'[20]

It is difficult to imagine any designer, even a global warming sceptic, using language like Loewy's today, at least not without lacing it with references to design's moral obligations. And rather than expecting to be applauded for producing something new, designers now anticipate having to justify doing so, and being challenged to defend other aspects of the ecological impact of their work.

There has also been a revision of design history. Buckminster Fuller is now regarded not just as an endearingly eccentric maverick, but as a gifted visionary, who anticipated many of design's current concerns; while the work of more conventional designers from the past has been reinterpreted to highlight its sustainable qualities.[21] For decades, the circular birchwood stool Model No. 60 designed in the early 1930s by Finnair's tardy celebrity passenger, Alvar Aalto, has been prized for its aesthetic merits as an exemplar of Scandinavian modernism.[22] The stool is still seen in that light but is also hailed as an early model of sustainability and locavorism, having always been made in exactly the same way by Artek, the furniture maker founded by Aalto, in the same factory, from the wood of silver birch trees grown in the same nearby forest. If a leg breaks, it should be possible to replace it with one from any other stool, regardless of when it was made. Similarly, one of Aalto's contemporaries, the Dutch designer and architect Thomas Gerrit Rietveld, has long been admired for the clarity with which he translated the geometric aesthetic of the De Stijl movement into objects like his wooden Red/Blue Chair. Now he is also praised for the use of found materials in the Crate Chair, which he made from the rough planks of spruce wood used in packing crates.[23]

Unlike Bucky, neither Aalto nor Rietveld styled himself as a sustainable pioneer. But as awareness of ecological issues grew in the second half of the twentieth century, spurred by the publication of books such as Aldo Leopold's *A Sand County Almanac* in 1949 and Rachel Carson's *Silent Spring* in 1962, a new cadre of designers emerged who were dedicated to environmental concerns.[24] An alternative design community emerged in northern California at the turn of the 1970s, when designer-makers moved there to live alongside fellow hippies in the redwood forests. Among them was James Blain Blunk, known as J.B., a physicist turned potter, who carved entire chairs and benches from enormous chunks of redwood and cypress found near his home beside a nature reserve in Inverness.[25] By the mid 1970s, the German design theorist Jochen Gros had left his job as a design engineer for Siemens to teach at the design school in Offenbach, where he founded Des-in, a group of design activists who experimented with the use of recycled materials and new forms of sustainable production.[26]

Another escapee from commercial design to academia, Victor Papanek, introduced the principles of ethically and environmentally responsible design to a wider audience in his 1970 book, *Design for the Real World*. Having grown up in the 'Red Vienna' that hatched the Isotype project, he fled Austria with his family when the Nazis gained power, finally settling in the United States. After studying architecture at Frank Lloyd Wright's school in Taliesin West, Arizona, Papanek worked in commercial design but loathed it, and devoted his working life to championing sustainable design in books, lectures and anthropological research projects conducted while living among the Navajos and Inuit and other indigenous communities. The design establishment detested him, which is hardly surprising, as the opening line of *Design for the Real World* was: 'There are professions more harmful than industrial design, but only a very few of them.' By the end of the page, he had also accused them of 'concocting the tawdry idiocies hawked by advertisers', putting 'murder on a mass-production basis' and of being 'a dangerous breed'. Four years after its publication, the British magazine

Design described Papanek as being 'disliked, even loathed by his contemporaries', and he later complained that they had 'derided, made fun of, or savagely attacked' his book.[27] But Papanek had the last laugh. By the time the second edition of *Design in the Real World* appeared in 1985, it had been translated into more than twenty different languages. Still in print today, it is one of the best-selling design books ever published.

In Papanek, Gros, Blunk and Bucky, the sustainable design camp is blessed with gutsy, charismatic idealists as role models, whose contribution to the wider history of design is now given greater acknowledgement. A lively community of websites, blogs, social media groups and professional networks has sprung up to champion their work and that of their successors, including Worldchanging, Inhabitat, Core 77, Good, Treehugger, Change Observer and The Designers Accord. The writing and lectures of the Canadian graphic designer Bruce Mau have played an important part in rooting environmentally responsible design within a cultural and political context, as have those of the British design strategist John Thackara and his Italian counterpart Ezio Manzini.[28] It has also been championed by the Danish government's €500,000 INDEX: Design to Improve Life award, and the Buckminster Fuller Challenge in which Bucky's charitable foundation gives $100,000 a year to a project that, true to his spirit, seeks to 'solve humanity's most pressing problems'.[29]

Such initiatives have fostered a receptive climate for sustainable design ventures, many of which are models of the new approach to design, not only in terms of their ecological objectives, but because they are fired by entrepreneurialism, rooted in digital technology and instinctively combine conventional design techniques with design thinking. As the Herculean challenge of redesigning the road transport system demonstrates, the problems facing sustainable design are so intensely complex, and often strategic and systemic in nature, that they demand the open-mindedness and ingenuity of a fluid, consensual tool like design thinking. Not that it is only of use on a large scale, as its role in smaller enterprises like the Daily Dump waste management programme in Bangalore illustrates.

Daily Dump was developed by the Indian designer Poonam Bir Kasturi as a fun, user-friendly way of encouraging people to recycle domestic waste and to compost it.[30] Once, cities exuded perverse pride in disgorging huge quantities of rubbish. The more debris they generated, the more prosperous they were likely to be. One of the most useful, yet least salubrious functions of design has been to find ways of disposing of trash so that the rest of us could forget about it. Now that we can no longer pretend to ignore the damage caused by doing so, getting rid of it responsibly has become a mark of civic sophistication. Very few cities have risen to the challenge, plunging refuse systems into crisis and causing acute problems, especially in the frenzied metropolises of rapidly expanding economies like India's. Bangalore is typical. It produces more than three thousand tonnes of waste every day and, as the government's composting plant can only handle five hundred tonnes of it, most of the rest is dumped illegally.[31]

Having chosen waste management as her cause, Bir Kasturi spent two years researching the best approach, funding the work with her savings. Her goal was to design a convenient way for people to recycle their refuse and to convert organic wet waste into compost. Working with Indian craftsmen, she developed terracotta pots for home composting, and produced a series of larger plastic composting tanks, all developed on an open source basis. She then set up a website to sell them.[32] Some people simply buy and use the composters. Others sign up for service packages, which include regular visits by the Daily Dump team to clean and empty the composters, and to do running repairs.[33] But Bir Kasturi has been more ambitious for Daily Dump. From the start, she wanted to add educational and entrepreneurial elements to the project, and deployed design thinking to do so. The website provides general information on recycling and composting, as well as about Daily Dump's products. It also encourages people to comment on the system, and to exchange tips with fellow users. The entrepreneurial opportunities can be as simple as selling compost and composters, or as ambitious as setting up 'Clones' of Daily Dump. Anyone wanting to do so registers their intention on the website, before downloading the design specifications of the composters for free. They are then encouraged to share their experience of running a Clone, to alert their peers to possible difficulties, and to flush out suggestions for improvement.[34]

Equally enterprising is FARM:shop in the Dalston area of east London, where the three founders of the eco-social design group Something & Son have transformed a derelict nineteenth-century terraced house, last used as a women's refuge, into a laboratory in which they are trying to grow as many different types of food as possible.[35] The three founders – Paul Smyth, a design engineer-turned-environmental activist, Andy Merritt, a graphic designer-turned-artist and Sam Henderson, a social scientist-turned-farmer – met while working on ecological projects in London, and hatched the idea for FARM:shop when the local council invited artists and designers to come up with interesting ways of using empty shops.[36] They were given a three-year lease and a renovation grant of £6,000. The building was in a dreadful state, with bars on the windows, but a friend tweeted about their plans on day one, and forty volunteers offered to help. After cleaning up the building, Something & Son contacted the manufacturers of the growing technologies they needed to cultivate food there, and worked with them to develop bespoke systems for the space.

Months later, one room on the ground floor had been transformed into an aquaponic farm for tilapia fish and lettuce. It was designed as a self-sustaining 'closed loop' system, with the waste from the fish filling the water with nutrients, which nourish the lettuce when the water is pumped into their tanks. The same water is then cleaned by the lettuce before returning to the fish tanks, and back again. Tomatoes, peppers, squash, basil and other plants are cultivated by hydroponic and aeroponic technology in what were once bedrooms. The basement is reserved for mushrooms, and more produce grown in a polytunnel in the backyard. Up on the roof, hens strut around a coop, seemingly oblivious to the noise and fumes of the traffic on the busy road outside. People wander in and out of the café in the room next to the aquaponic tanks, where they tuck into the produce grown at FARM:shop, and on an organic farm an hour's drive away where Henderson is based. There have been crises, such as bugs infesting tomatoes, spider mites attacking luffa sponges, the mushroom crop failing and a power cut that threatened to kill the fish. (As tilapia are Egyptian, they can survive only in heated water. When the power failed, the temperature of the water plummeted, but thankfully the electricity came back on before any damage was done.) Despite these setbacks the experiment worked, and proved that Something & Son could develop a financially sustainable model of growing food in such cramped urban conditions.[37]

Like Poonam Bir Kasturi, Something & Son have had the freedom to work on their own terms. Other designers who share their objectives may be more constrained. They might work for a commercial design consultancy which has been contracted to advise clients on an ad hoc basis, as IDEO was by Oral-B. Or they could be employed in the corporate design team of a business that gives them little or no influence over many of the decisions which will determine the environmental impact of their work. A sustainable purist would not tolerate such compromises, but not all designers are able or willing to extricate themselves from them – though such constraints may lessen in future as another positive development in sustainable design is the sea change in corporate thinking.

Once regarded as the preserve of hippy entrepreneurs and cranky executives with personal passions for ecology, sustainability is now seen very differently in corporate circles, thanks to the pioneering companies that have proved its financial value. Among them is Interface, the American carpet tile manufacturer, which was not the likeliest of sustainable champions, not least because its founder, Ray Anderson, had no discernible interest in the cause.[38] Having established Interface as the world's biggest company in its field, he was the epitome of an ambitious, profit-oriented American businessman. When his former colleagues had suggested that it might be a mistake to leave a well-paid job in an established carpet company to set up on his own, he rebuffed them with: 'The hell you say.' Interface became powerful and profitable in the traditional way by producing carpet tiles with no apparent regard for the ecological consequences. It used lots of petroleum-based materials including nylon, polyester and acrylic, as well as toxic dyes and glues, and its smokestack factories belched dark clouds of pollution up into the sky.[39]

When a client suggested in the mid 1990s that Interface needed an environmental strategy, Anderson was far from enthusiastic, but he swotted up on the subject by reading *The Ecology of Commerce*, a book by the environmentalist Paul Hawken. One section of the book describes how twenty-nine reindeer were imported to a US government research station on St Matthew Island in the Bering Sea in Alaska in 1949 to provide an emergency supply of food. The station closed a few years later, leaving the reindeer behind to breed freely. By 1963, there were six thousand of them, but two years later, only forty-two were still alive. The others had starved to death, having stripped the island of vegetation.[40] Anderson was deeply moved by the story, which he saw as a metaphor for the self-destructive way that companies like his were depriving the earth of natural resources. He astonished his employees by announcing that Interface was to become what he called a 'restorative enterprise', with the aim of having no negative impact on the environment by

2020, producing no greenhouse emissions and no waste. 'I have to admit I thought he'd gone around the bend,' said one colleague, Dan Hendrix, who was eventually won round and became chief executive, but Anderson was undeterred.[41]

Every aspect of Interface was scrutinized to ascertain how it could be made more sustainable. Renewable energy was used wherever possible, and carbohydrate polymers developed to replace petrochemicals in the carpet tiles. Abandoned carpets were recovered from landfill sites for recycling, and customers encouraged to return their old carpets for the same purpose. Quality control was improved to reduce waste. Interface invested heavily in developing the new production technologies required to achieve its objectives, many of which did not exist. Anderson and his team combined conventional approaches to industrial design with design thinking, often doing so instinctively as Josiah Wedgwood and his colleagues had done in the vanguard of the first wave of industrialization in the late eighteenth century.

By 2007, twelve years after he had read Hawken's book, Anderson claimed that the company was halfway up 'Mount Sustainability'. (He had a soft spot for flowery corporate jargon.) The use of fossil fuels had almost halved, and the factories consumed a third as much water. Some seventy-four tonnes of discarded carpet had been recovered from landfill sites and Interface's own contribution to landfill was down by eighty per cent. Anderson calculated that the company was saving over $300 million a year from quality improvements, more than enough to pay for the research and development required. Moreover, sales had risen by two-thirds since the start of Mission Zero, and profits had doubled.[42] 'What started out as the right thing to do quickly became the smart thing,' said Anderson, who took to calling himself 'a recovering plunderer'[43] while insisting that he remained 'as profit-minded and competitive as anyone you're likely to meet.'[44] By the time he died in 2011, the carbon footprint of Interface's factories had halved, and he was hailed by the *Economist* as 'America's greenest businessman'.[45]

Other companies have learnt from Interface's example that a sustainable business model can produce, as Anderson put it, 'not just bigger profits, but better, more legitimate ones too'. They have discovered that they too can save money by using water and energy more economically, sourcing materials more responsibly and planning transportation more efficiently. Operating sustainably can also boost their share prices by attracting new investors among the growing number of ethical funds, which only invest in businesses that meet certain environmental and ethical standards. Another advantage is existing employees are inspired by the corporate goals, which help companies to attract talented new recruits, such as sought-after graduates, who can take their pick of prospective employers. All compelling reasons for businesses to rethink the way they work, what they produce and how it is designed, and for other organizations to do the same. Not that sustainable design is any less challenging, but the growing confidence in its benefits should foster more models of good practice in future, whether they are corporate giants like Interface, dynamic entrepreneurial enterprises akin to Daily Dump, or cheeky examples of environmental design activism such as FARM:shop.

11 Why form no longer follows function

It is not enough for designers today to balance form and function, and it is also not enough simply to ascribe meaning. Design now must imagine all its previous tasks in a dynamic, animated context. Things may communicate with people, but designers write the initial script that lets us develop and improvise the dialogue.
— Paola Antonelli[1]

Form follows function. Those three words have been trotted out so often that, as well as becoming a cliché, they have acquired their own mythology. Not only are they generally misquoted – there should be four words, not three, because the correct wording is 'form ever follows function' – they are routinely misattributed, typically to a modernist grandee such as Le Corbusier or Mies van der Rohe, when the phrase was actually coined by the less well-known American architect Louis Sullivan.[2] He wrote it at the end of a very long sentence in an 1896 essay entitled 'The Tall Office Building Artistically Considered': 'It is the pervading law of all things organic, and inorganic, of all things physical and metaphysical, of all things human and all things super-human, and all true manifestations of the head, of the heart, of the soul, that life is recognizable in its expression, that form ever follows function.'[3]

What Sullivan meant was that the size, style and structure of a man-made creation, whether it is a tiny object or a sky-scrapingly tall building, should be defined by its purpose. In other words, how the thing looks should be dictated by what it is intended to do. Think of a spoon, a chair or any other object which is a pure form with no motor or microchips to power it. Even if you were seeing one for the first time, you could probably guess roughly what to do with it based on its shape, size and solidity. Eventually, you should recognize that the chair offers some sort of support to the human body, and that the spoon is intended to insert something into it, because the form of those objects generally follows their function.

The same applies to old-fashioned electrical products,

Something & Son's food growing experiment in Dalston, London

OPPORTUNITY
NEW RETAIL LEISURE
RESTAURANT UNITS

CINEMA

such as television sets or radios. Doubtless, it would take you longer to figure out what to do with them, but there would be some clues. Does the fact that images flash across the biggest part of the television suggest that it is meant to be looked at? Given that sound booms out from the largest area of the radio, is it something to be listened to? In both cases, your judgement will have been based mostly on how those machines look, just as it was for the pure forms. But Sullivan's principle has no relevance for digital devices, like the inscrutable smartphone at the beginning of this book, and its demise has transformed the type of objects we use every day, and how we relate to them.

When Sullivan decreed his law in the late 1890s, it synched perfectly with the scientific theory of natural selection, which had dominated the intellectual agenda since the publication of Charles Darwin's *The Origin of Species* nearly forty years before.[4] To those who believed in the Darwinian concept of 'the survival of the fittest' in nature, it seemed logical that it should apply to industry too. If a zebra's stripes acted as camouflage to help hide it from potential predators, and a giraffe's long neck enabled it to forage for food at the top of trees by reaching higher up than other mammals, surely factory-made products could also derive practical benefits from elements of their appearance? At the time Sullivan wrote his essay, the idea that the artefacts of the industrial age might have a close rapport with nature would also have seemed comforting, for the same reason that the gentle curves and gleaming wood of Thonet's bentwood chairs must have felt reassuring because of their resemblance to traditional rustic furniture. Such allusions helped to make industrialization appear less intimidating then and have continued to do so ever since, not least as there are so many compelling examples of natural design ingenuity. An eggshell, for example. Not only is it very beautiful to look at, but like a banana skin it is robust, user-friendly and biodegradable. The shell is easily broken when you wish to eat its contents, and just as easily disposed of. How could any form of man-made packaging be expected to improve upon it?

The more overtly industrial a product, the greater the need for reassurance. Take the aeroplane. Even for those of us who are lucky enough not to be flight-phobic, it requires a huge leap of faith to walk into a long, skinny container in the confident expectation that it will leap off the ground and whisk you thousands of miles through the air before landing safely; so huge a leap that it is best not to dwell on it. Imagine how much more terrifying such a prospect would have seemed when aeroplanes were new. Even the most nervous first-time fliers may have felt slightly less frightened because those newfangled flying machines were made in the same shape as birds, creatures that they knew for sure could fly safely, having seen them doing so repeatedly. Had they ploughed through theories of aerodynamics, those *ingénu* air travellers would also have discovered that, true to Darwinian principles, the best-designed shape for anything intended to fly is indeed that of a bird. Decades later, when the enormous Boeing 747 was launched in 1970, nearly three-quarters of a century after Sullivan had coined his famous phrase, people still sought comfort in the notion that air travel was not as unnatural as they feared by nicknaming it 'Big Bird' and 'Jumbo Jet'.[5]

As well as by Darwinism, Sullivan's law was affirmed by another belief system: modernism. As the modern movement gathered force in the early twentieth century, the idea that the physical form of the products which were rolling out of factories should be dictated by the need to fulfil their practical purpose not only seemed appealing but synched with aesthetic trends. After centuries in which intricate handcraftsmanship had been considered the acme of 'good taste', the 'machine age' modernists favoured a studiedly simple design style that used minimal materials, ideally with no decoration, lest it detract from the object's performance or its sense of purpose. Not for nothing was another modernist cliché, one that really can be attributed to Mies, 'less is more'.[6]

Similar principles had been applied for centuries to the anonymous design of utilitarian products such as hammers, chisels, planes and other tools, and still are today. Take one of Fiskars' axes. A film on the company's website explains how its design has been finessed to optimize efficiency, even when the axe is used intensively in extreme weather. To prevent it from slipping, the handle has been made from a special rubber and tapered into a hook at the end. The edge of the blade has been bevelled to form the best possible angle to enhance the axe's chopping power, and to render it easier to remove the blade from blocks of wood; the head has been moulded securely on to the handle to stop it from shaking loose, and so on.[7] Fiskars' tools apart, such workmanlike objects are rarely praised for their design qualities, yet the deployment of 'form follows function' produced some of the most admired examples of twentieth-century design.

Consider cutlery. My own favourite knives, forks and

spoons are the ones designed by the Danish architect Arne Jacobsen in the late 1950s for the SAS Royal Hotel. Each one is a spindly slither of stainless steel, shaped very differently from conventional cutlery. Jacobsen began by analysing how the different pieces would be used, and made them from the minimum amount of metal required to do the job. The end results nestle neatly in the palm of the hand with wide, smooth surfaces for the fingertips to grip, narrowing at the point where they enter the mouth. Jacobsen allowed for functional details that had been ignored by his predecessors, such as designing asymmetrical soup spoons in an attempt to prevent slops and spills.[8] By connecting the handle of each one to the top of the end-piece rather than to its centre, he ensured that if the spoon slips, the soup falls away from you. Jacobsen even produced different versions of the soup spoon for left-handed and right-handed people, the handle extending to the left for the former and to the right for the latter. He also refined the shape and size of the end-piece so that it delivers a perfectly judged quantity of soup to the mouth and cools it slightly beforehand without losing all of its heat.[9]

Such subtleties were lost on the Danish media, which poked fun at Jacobsen's strangely shaped cutlery when the hotel opened in 1960. One newspaper dispatched a journalist to the restaurant, where he was photographed struggling to eat peas on a skinny fork. Eventually, the hotel replaced Jacobsen's designs with a conventional style of cutlery,[10] yet the originals have sold steadily ever since. When Stanley Kubrick was choosing suitably futuristic props for his 1968 sci-fi epic *2001: A Space Odyssey*, he commissioned most of them to be designed from scratch, but made an exception for the cutlery. Kubrick considered that Jacobsen's ten-year-old exercise in 'form follows function' could pass plausibly for something that belonged thirty years into the future. On a practical note, I have used mine every day for over twenty years.

'Form follows function' has influenced the design of so many of the things around us that many of our instinctive assumptions about our surroundings are rooted in it. When confronted by new objects, we are inclined to guess what to do with them by looking for physical similarities to familiar ones. During the 1950s, when Henry Dreyfuss's design firm was working on an airline project, he dispatched a team of researchers to quiz cabin crews on how passengers were responding to the existing cabin designs. A United Airlines stewardess reported that several people had 'posted' letters inside the air-conditioning slots, having mistaken them for mailboxes. And a TWA stewardess recounted how well-intentioned parents had deposited their babies in the overhead baggage racks, blithely thinking that they could slumber there safely throughout the flight. In both instances, the passengers had made what they thought were sensible assessments based on their past experience of using things that resembled those slots and racks, and had acted accordingly.[11]

When it came to developing electronic products, designers turned similar visual cues to their advantage. Not only are those objects more complex technically than 'pure forms' such as knives, forks, spoons and chairs, they are potentially trickier to operate, not least because the machinery that drives them is hidden from view inside their casings, and would be incomprehensible to most people if they saw it. Dieter Rams and the Braun design team were expert at making electronic gizmos seem straightforward even to people using them for the first time by nudging them towards instinctively doing the right things in the right sequence. When Rams spelt out his vision of good design in 'Ten Principles of Good Design', he devoted the sixth principle to the importance of usability. 'Good design helps to understand a product,' he stated. 'It clarifies the structure of the product. And more: the product speaks – in a sense. Optimally, the product is self-explanatory.'[12] In other words, he believed that if the form of an electrical device was dictated by its function, it would be easier to operate, as well as more efficient.

Helping people to work a 1960s record player or a 1970s juicer like the ones designed for Braun by Rams and his team was a doughty challenge, but their designers did, at least, have the possibility of creating physical prompts to guide them by adding controls such as buttons and switches. The designers of the next generation of products – digital devices – had no such luxury, because unlike spoons, chairs and aircraft, their appearance bears no visible relation to what they do, due to the impact of Moore's law on their ever-increasing power and shrinking size.[13]

The computer is an example. Its design history dates back to the 1820s when the British mathematician Charles Babbage built a machine that could complete advanced mathematical calculations faster than the human brain. He then refined it, helped by his friend Ada Byron. The daughter of the poet Lord Byron and a mathematical prodigy as a child, Ada was happily married to the Earl of Lovelace, yet devoted to Babbage. James Gleick told their story beautifully in his book *The Information*.[14] Over a century after Babbage's first invention, Alan Turing conceived the modern computer when he described an electronic machine which would be versatile enough to perform numerous different tasks by changing its software rather than its hardware in a 1936 research paper, 'On Computable Numbers'.[15] Turing continued his research while working at the government codebreaking centre at Bletchley Park in Buckinghamshire during the Second World War, when Britain, like other countries, stepped up its investment in computing as part of the defence effort. The first stored program computer, akin to the one envisaged in his paper, was 'Baby', which was completed at the University of Manchester in 1948.[16] Turing moved there months later and worked on the project until his tragic death.[17]

Mathematicians and scientists were engaged in similar exercises in other countries, including a team of researchers led by John von Neumann at Princeton University in New Jersey. Using funding provided by the US government as part of the hydrogen bomb project, they developed the fastest computer of the era in 1953. Von Neumann had persuaded the authorities that it would be able to conduct theoretical tests of the bomb.[18] By deciding against patenting the machine and by sharing their research findings with their peers, his team ensured that their design was widely copied, notably by IBM in the development of its first electronic computer, the 701.[19] IBM tidied up its machine by encasing the components in metal cabinets, but the original computers at Manchester and Princeton were labyrinths of tangled wires and dials. By the time the transistor was invented in the late 1950s, the machines had become neater but were rarely seen outside scientific laboratories or university research centres.

The first computer I ever saw was in a factory near Brussels which I visited with my father in 1965. Dad was a mechanical engineer, who was designing a production plant there, and he had taken my mother, brother and me with him on a business trip. I was six years old, and remember the trip for three things: eating chips smothered in mayonnaise, Belgian-style; the dazzling sight of the Atomium, the futuristic tower built as the centrepiece of the 1958 Brussels World Fair; and that first glimpse of a computer.

It was enormous, so much so that it was not one machine but an entire room filled with machinery, the 'Computer Room'.

My memory is of seeing lots of metal doors and, although I may be imagining it, hearing a loud clanking noise. Dad explained how lucky we were to have been allowed to enter the room and to see its contents, because the machines hidden behind those doors were very special, and so expensive that few companies could afford to buy them. I was impressed, and so I should have been. A system like that would have cost a fortune, and required a dedicated team of technicians to maintain it. The mainframe computer inside needed to be kept in an air-conditioned space, which was locked at all times. In the mid 1960s, even senior executives were not permitted to enter without special permission and had to ask the technicians to operate the machines for them. Thrilling though the scale and secrecy made it seem, all of the kit I saw in the Computer Room in 1965 may have had less power than a laptop today.

Over the next decade, computers became slightly more powerful, a little less enormous and expensive, but were still sealed inside air-conditioned rooms. In 1975, IBM introduced its first 'portable computer', the 5100, which was small enough to sit on a desktop, but only just portable, given that it weighed fifty pounds. Hefty though that sounds, an equivalent machine would have weighed half a ton five years before, and taken up the space of two desks. But the 5100 had its drawbacks. One was the price, which ranged from $8,975 to $19,975. Another was its complexity: only proficient technicians could use it.[20]

By the late 1970s, a group of geeky hobbyists and inventors, including Steve Jobs and Steve Wozniak in the United States, and Clive Sinclair and Adam Osborne in Britain, were developing smaller, cheaper machines, initially sold in kit form to fellow obsessives.[21] Their design challenge was to work out how to enable everyone else to use them too. Up until then, computers had been operated using code, which was baffling to anyone who was not a trained programmer. Their solution was to replace the coded instructions to the computer with cues, such as physical gestures or visual symbols, which would make sense to amateurs.

The typewriter keyboard was already established as the principal means of relaying information to the machine, as was the mouse (so called because it was a similar shape to a mouse's body with the cord as its 'tail') as a device for identifying symbols on the screen, and instructing the computer to move or delete them.[22] Presenting the other control mechanisms in an accessible form proved elusive until Tim Mott, a researcher at the Xerox Palo Alto Research Center, or Xerox PARC, in California, realized that there were similarities between the way people organized their office paperwork and how they might manage information on a computer screen. Having hit upon the idea while waiting for a friend in a bar, he sketched it on a napkin and showed it to a colleague, Larry Tesler, who helped him to develop it. 'It was a set of icons for a filing cabinet, and a copier, or a printer in this case, and a trash can,' Mott recalled. 'The metaphor was that entire documents could be grabbed by the mouse and moved around on the screen. We didn't think of it as a desktop, we thought about it as moving these documents around an office. They could be dropped into a filing cabinet, or they could be dropped on to a printer, or they could be dropped into a trash can.'[23]

The 'office' and other early ideas for graphic interfaces – the technical term for the presentation of information on computer screens – developed at Xerox PARC were commercialized by Apple and Microsoft in their software, and have influenced the way we use computers ever since. 'The main thing in our design is to make things intuitively obvious,' Steve Jobs explained in a speech to the International Design Conference in Aspen in 1983. 'People know how to deal with a desktop intuitively. If you walk into an office, there are papers on the desk. The one on the top is the most important. People know how to switch priority. Part of the reason we model our computers on metaphors like the desktop is that we can leverage the experience people already have.'[24]

Bill Moggridge's book *Designing Interactions* charts the work of the designers and programmers that developed those metaphors, including Mott, Tesler and Bill Atkinson, who was responsible for many of Apple's early interfaces, such as the first Macintosh and the Lisa. While working on the Lisa, Atkinson realized that they had to find a way of telling the user if the trash can needed to be emptied. His original proposal was that flies should buzz around the top, but his colleagues overruled him on the grounds that the sight of fetid insects might put people off. They suggested filling the digital trash can with crumpled sheets of pristine white paper instead.[25]

As computers have become ever smaller, yet faster and more powerful, their designers have searched for new ways of making those opaque plastic boxes easier to use, or to look as though they are. That is why Apple made sure that the first things anyone saw on the screen after switching on the 1984

Macintosh were the words 'Hello' and 'Welcome'. It also explains why the company spent $750,000, a record for the time, on producing a television commercial directed by Ridley Scott that introduced the Macintosh by subverting the dystopian expectations of the year 1984 fostered by George Orwell's novel of the same name. The slogan was: 'Why 1984 won't be like "1984"'.[26] The quest for usability then influenced future design decisions, such as exposing the innards of the first iMac in 1998, as well as cheering up the case with colourful plastic and encouraging people to touch it by popping a handle on top.

Similar principles have since been applied with varying degrees of success to other digital devices, whose size and power has been transformed by advances in integrated circuit technology, and in other components, such as batteries. Early mobile phones were nicknamed 'bricks', because their batteries not only were shaped like bricks, but weighed almost as much. Anyone who tried using one of those phones will remember how heavy they were and how quickly the batteries died. My first mobile in the early 1990s was so hefty that my wrist ached after holding it for more than a few minutes. The battery was to blame, and it only lasted for half an hour before conking out. Phones now function for days on tiny batteries, and are a fraction of the size of the original 'bricks'. Other industries have benefited too. For many years, a major obstacle to the development of electric cars was finding batteries which were compact enough to fit inside the vehicle, yet powerful enough to drive it. Thanks to the telecommunications industry's investment in battery development for the immensely profitable phone market, suitable batteries became available and have been used to power hundreds of thousands of electric cars and trucks.

Equally important have been the leaps in wireless technology that enable tiny digital devices to access powerful computers. One reason why the iPod can be so small, and seem so straightforward, is because most of its functions are performed by Apple's gargantuan computer system, to which it is connected wirelessly. The same applies to digital and electronic readers, and any other devices that offer the possibility of accessing huge quantities of information.

The more complicated digital products have become, the more important it is for using them to feel simple. Our experience of operating them has become as critical in determining how we feel about the objects, and whether we deem them to be well designed, as traditional concerns like how they look and what they do. The technical term is 'user interface design', or UI for short, and its goal is to enable us to use complex devices so intuitively that we never need to puzzle over what to do next. Whether you are using your tiny phone to make a call, or to execute one of the hundreds of other tasks of which it is capable, the challenge for the interface designer is to ensure that you can do so effortlessly.

Achieving this is a tortuous process of trial and error, which often seems to be a thankless task. The best examples of user interface design are the ones that feel simple, and because they do, we are unlikely to notice their merits or appreciate their sophistication. Even if we did, few of us are technologically adept enough to pinpoint exactly what the designers have done, or how they did it. Nor do we have an adequate vocabulary to describe the strengths of a product's user interface or its innovative qualities. Apple is regularly praised for designing user-friendly products, yet even its most ardent admirers rarely manage to say more than that they seem 'intuitive', 'simple' or 'uncomplicated'.

But when user interface design goes wrong, we notice immediately, because it can be disruptive, sometimes hellishly so. How many times have you bought a new phone, usually a snazzier one, only to discover that you need to learn how to use it all over again, even if it is the same brand? Some of your favourite features may have disappeared, and executing the same function might require even more scrolling and button pressing than it did on the old phone. Typically we blame ourselves for being technically inept when such things happen, but we are often wrong to do so, because the irritating fiddliness of that phone is probably down to lazy user interface design. So is the tendency for 'Come in, Cape Canaveral' television remote controllers to sport more dials and buttons than Kennedy Space Station, and the Herculean challenge of programming a digital television set.

Once I went to dinner at a friend's house where the bathrooms are fitted with digital taps. It was easy enough to turn those taps on, but adjusting the temperature of the water appeared to require a doctorate in advanced computer programming and, try as I might, I could not work out how to turn them off. Lights flickered, as did various digital symbols, but the water gushed relentlessly. I called for help, but my friend was stumped too. Eventually she went online to consult the help section of the manufacturer's website. Another friend was forced to resort to desperate measures when his digital home-management system failed. He returned home one night to discover that the lighting, computers, phones, television screens, kitchen appliances and all the other digital systems were flashing on and off repeatedly. Nothing he tried on the control panel worked. After several hours of bedlam, he picked up a hammer and smashed the control panel. Peace at last. Incompetent user interface design was not the cause of the problem, but it impeded his efforts to find a solution.

Why is so much user interface design so bad? One explanation is that many 'new' interfaces are really upgrades of existing ones to which extra features have been added. Phones are prone to this. Often they seem inexplicably over-complicated because, ideally, their operating systems should have been redesigned from scratch, rather than hastily augmented. In this respect, they are akin to houses which have acquired various additions and extensions over the years. Somehow, they seldom appear as coherent as a new house which was designed and built to accommodate them all.

But most user interfaces fail because of laziness, ineptitude, thoughtlessness, inadequate testing, lack of foresight and the other failings that bedevil so many areas of design. Logic suggests that if, say, a new computer operating system is to be easy to use, the most popular commands should be given the most prominent positions on the screen. All too often, the largest and most conspicuous commands are not the most useful ones. If, for example, 'Copy' and 'Paste' are likelier to be used more frequently than 'New folder', why not locate them accordingly to make them easier to identify? The same should apply to the relative positioning of every other command. Unless such nuances are reflected in the interface's design, it may be unnecessarily difficult for people to operate the system, although few of them will ever realize why.

Despite all these shortcomings (and it is rare to find a technologist who does not complain about sloppy user interface design) a measure of its success is the speed with which we have adopted so many new digital devices and learnt how to use them. A decade ago, when Tom Cruise's character, Chief John Anderton, barked instructions or waved imperiously towards giant computer screens in Steven Spielberg's sci-fi thriller *Minority Report*,[27] those technologies seemed seductively futuristic; now they are ubiquitous, as we can control digital devices by issuing vocal commands or making physical gestures, rather than through intermediaries such as keyboards and keypads or by pointing at clusters of pixels with a mouse.

When I was a child, and my brother and I were watching a western on television, we would each jump on an arm of a sofa and 'ride' it like a cowboy's horse. I then spent much of my teens belting out 'hairbrush hit' versions of favourite songs into upturned hairbrushes-cum-microphones in girl groups with friends. Games systems like Microsoft's Kinect operate on the same principle by deploying motion sensors to respond to the movement of our bodies. Felling a virtual baddie with a perfectly executed scissor kick feels so much more exciting than doing so with a controller.[28] Many of the ways we use our smartphones have the same elemental appeal. Think of how much more pleasurable it feels to zoom in towards a detail in a photo by stroking a phone screen with your fingers, rather than doing so remotely by pressing a button. Unfathomable though the technology is, those physical gestures feel so simple and natural that they imbue the experience with those qualities.

Similarly, personal monitoring devices use the tiny motion sensors in a wristband to monitor your movements while you are awake and asleep, to tell you if you are eating healthily, taking enough exercise and sleeping satisfactorily. The data collected by the sensors is dispatched to an application which tots up how many calories you have burnt, say, by walking to work or climbing the stairs, and whether you slept well. If you take photos of your food on your phone, the app analyses the calorie content. The wristband can also be programmed to act as an alarm, which wakes you by vibrating at the ideal moment in your sleep cycle. It can even tell you when you have been slouching in a chair for too long.[29] And as brain sensor technology advances, we will be able to control other devices with our thoughts, by playing video games, for instance, on a system that can detect our wishes without our needing to express them physically or verbally.

All this may seem a long way away from Sullivan's law in the 1890s, but as soon as form no longer followed function, designers needed to find new ways of enabling us to communicate with objects and vice versa. Our expectations of design and, specifically, the aspects of it that we value the most, are changing accordingly. One outcome is the increasing importance of ease of use in design, especially for anything digital. Another is the obsession with simplicity, or the illusion of it. And a third is the recognition that many of the familiar products which we have used for years, but whose functions can now be executed by a smartphone, will soon disappear.

The question is: which ones? The answer is likely to be determined by the underlying principles of another Darwinian phenomenon, 'sexual selection', or, as the American philosopher David Rothenberg calls it, 'aesthetic selection' and 'the survival of the beautiful'.[30] In his 1871 book *The Descent of Man*, Charles Darwin sought to explain why the males of certain species appear to defy the law of natural selection by retaining some features that serve no practical function. An example is the peacock's splendid tail, whose sole purpose, Darwin argued, was to be so beautiful that it would arouse the sexual interest of females of the species, in this instance, peahens, thereby enabling the male to breed and his species to thrive.[31] A similar argument can be applied to the evolution of products in the digital age. If the traditional form of an object is exceptionally pleasing to look at or to touch, it is likely to survive because we will continue to crave it, even though a more convenient digital alternative is readily available. Beautifully bound books are prime candidates. They fulfil their practical purpose by protecting the words, while providing a compact, portable form in which to read them, and can also give us something that has so far eluded electronic and digital books: the sensual pleasure of seeing exquisite typography or illustrations, and of smelling or touching fine paper. Whereas the prospects of purely practical gizmos like the pocket calculator – which once seemed so thrillingly innovative that it inspired Kraftwerk's 1981 single 'Pocket Calculator'[32] – are bleaker. Why bother buying one, when the calculator program on your phone will do the same job as well, if not better?

Last but not least, another legacy of the toppling of 'form follows function' is a subtle transformation in our perceptions of size and its perceived value. For centuries, although 'big' was not necessarily always seen as being 'best', it is fair to say that it was generally expected to be 'better', because the biggest things were typically also the most powerful, luxurious and expensive. The faster the car, the larger its engine tends to be; and the most coveted room in a hotel is still generally the one with the most space. Even the exceptions have often proved the rule. Anyone who flew on Concorde will remember how odd it seemed that the cabin of the world's fastest commercial aircraft should be so small. Crass though it sounds, there was a comforting psychological subtext to the 'big is generally better' principle of literally being able to see that you really were getting more for your money.

But since Moore's law took effect and computing power has been compressed into ever tinier transistors, the biggest things have no longer necessarily been the most powerful, and that role has been commandeered by some of the smallest objects we own. As John Maeda put it in his book *The Laws of Simplicity*: 'The computational power of a machine that sixty years ago weighed sixty thousand pounds and occupied one thousand eight hundred square feet can now be packed onto a sliver of metal less than a tenth of the size of the nail on your pinkie.'[33] An inevitable consequence of so dramatic a reversal is that our value judgements about size have changed.

The demise of Sullivan's arcanely worded law has had such a profound impact on the contents of our lives that the Apple design team hailed it with a rare insider joke when the second version of the iPod shuffle was introduced in 2006. It consists of a brushed aluminium box which weighs an ounce and is half a cubic inch in size with a clip on the back.[34] As you would expect of an opaque digital device, nothing in its appearance suggests what it might do, the only visible functional component being the clip, which attaches the shuffle to a belt or lapel. That was the punchline. When it comes to the clip, the form of the shuffle does indeed follow its function, but that particular function is arguably the least interesting thing it does.

12 Me, myself and I

> **Life goes on here without fading or flowering, or rather death never ends. The many kinds of objects differ only in their rhythm of attrition or renewal, not in their fate. A pencil lasts a week; then it has run its course and is replaced by an identical one . . . The rush-seat chair is allotted three years until it is due to be replaced, the individual who sits out his life on it some thirty or thirty-five years of service; then a new individual is seated on the chair, just the same as the old one.**
> **— Stefan Zweig[1]**

Do you ring a doorbell with a finger or a thumb? The answer will reveal your age almost as accurately as the way you dance or how wrinkly your hands are. The older you are, the likelier you will be to press it with a finger, probably your index finger. If you are younger, you may well use a thumb, because it will have been exercised so thoroughly by typing text messages and gunning down digital assailants on game consoles that it is likely to be stronger and nimbler than any of your fingers.

Ringing a doorbell is one of those mundane actions to which we give little thought, but execute instinctively, as efficiently as possible. Making the unconscious choice to use a thumb, rather than a finger, demonstrates how changes in our designed environment can affect our behaviour. Another example is how many once-useful skills, often painfully acquired, have become, if not quite obsolete, no longer as valuable as they once were. Who needs a good sense of direction in the age of Google Maps and satellite navigation systems? An ability to spell now we have spellcheck programs? A talent for mental arithmetic when phones have calculator apps? And as for being good at inventing games, useful though that once was, these days there is *World of Warcraft* or *Angry Birds*.

Conversely, we have acquired new skills; so many that one of them is an aptitude for change. The contents of our daily lives have changed more dramatically since the Internet's invention in 1991 than at any time since the introduction of electricity at the turn of the twentieth century.[2] Think of the dozens of new technologies we have to learn to use every year. Controlling a computer by moving your hands once seemed thrillingly futuristic, but is now a standard component of video games. The same goes for passing through airport security without showing your passport because you can verify your identity by popping into a booth to scan your irises,[3] and for checking on to a flight using your smartphone rather than a printed boarding pass. So many technologies come and go that no sooner have we trained ourselves to use a new form of hardware or software than another appears and the learning process starts again.

We have become equally adept at multitasking. There was a time when being 'on the phone' served as both an excuse and an explanation as to why you could not be expected to do anything else. Now, it is a means of giving a running commentary on whatever else you are doing at the time on a plethora of screens and networks. Using several devices at once has enabled us to do various things simultaneously in other areas of our lives too. Not that it necessarily helps us to concentrate on each one, at least not as intensely as we might wish. When the novelist Zadie Smith cited her ten rules for writing, number seven was to 'work on a computer that is disconnected from the Internet'.[4]

Another new skill is synthesizing. Thanks to Moore's law, we are assailed with so much information that we have had to learn how to ignore the flotsam and spot the gems. We then assess its quality in terms of accuracy and objectivity. Some of the sources will be familiar and trustworthy, like a news app you read every day. Others will be unknown, such as bloggers you have not heard of and may never encounter again, or sources we know to be risky, like Wikipedia. The degree to which the information has been synthesized will vary too. It may have been meticulously researched, fact-checked and edited in the traditional way. Or you may have unearthed raw opinion that could turn out to be anything from scrupulously objective to the type of ragingly opinionated Twitter blasts that the American computer scientist Jaron Lanier described in his book *You Are Not a Gadget* as reducing personal tragedy and humanitarian disasters to a maximum of one hundred and forty characters.[5] How can you tell which is which? With difficulty, but we have all had to try, which is how we have learnt, however laboriously, to become sharper at synthesizing the information that is presented to us online and off. Our visual processing skills have improved for the same reasons, making us more proficient at analysing rapid streams of visual information and at identifying individual images, even ones we see briefly.[6]

As well as prompting us to learn new skills and to jettison old ones, our experience of the designed environment has changed our expectations. One is that we expect the pace of life to be faster. We are also more amenable to collaboration thanks to all the online petitions we have signed, and useful insights unearthed from collective endeavours like Wikipedia. And we are more questioning, which is not surprising given that all our synthesizing has taught us to be constructively sceptical, as has the knowledge that every image we see, still or moving, may have been digitally altered. Once, when people saw something astonishing, they would say: 'I can't believe my eyes.' Now, we do not expect to be able to. How could we, when we know that the movie star on a magazine cover may not be as slim as she appears, or that pimply teen stars can be 'cured' of acne by a few clicks of a mouse?

Significant though all these changes are, another shift in our expectations has even more radical implications for design: the desire to express our individuality. It is easy to understand why self-expression has become so important. Think of the way we use the Internet. If you and I typed the same question into the same search engine, we would be presented with an identical list of web entries, but would probably choose to explore different ones, in different sequences. Once we had checked out a particular entry, each of us might decide to flit off into another blog or website, doubtless different ones again. We may end up with the same answer, but we would be unlikely to have found it in the same way. Even if you personally put the same question to the same search engine on different days, the results could be equally diverse. The way in which you choose to analyse the available information would depend on your mood, the amount of time you had to spend on the search and how preoccupied you were with other matters. Every time we use the Internet, we determine our own idiosyncratic paths around it, picking a route, deviating from it at will and ending it at random, having pieced together the information bit by bit.

We determine our own destinies in video games too, and in the experience of conceiving, planning, designing and constructing virtual worlds in games like *The Sims* and *Spore*, or online environments such as Second Life and YoVille. We spin our life stories on Facebook, comment on them on Twitter and illustrate

Unfold's experiments with 3D printing technology

them on Pinterest. When we write an email, print a letter or read an ebook, we expect to choose which size and style of font to use, turning some of us into amateur type buffs. (Hence the outcry when IKEA abandoned the purist modernist font Futura in its corporate logo for the digital typeface Verdana.[7])If we do not like the look of a website, we can redesign it with an app like Readability, and do the same on an iPad with Instapaper.[8] And we are rapidly becoming accustomed to redesigning all – or parts – of ourselves. If you do not like a photograph of yourself, you can digitally erase the offending features, or you can achieve similar effects for real with cosmetic surgery, prostheses, hair dye, wigs, coloured contact lenses, make-up and other extreme, and not so extreme, 'beauty' treatments.

The outcome is that we are increasingly eager to take design decisions ourselves, rather than delegating them to designers. Not that personalization is new. Wealthy people have always been able to pay for things to be designed and made specially for them, as Qin Shihuangdi did by orchestrating his fantastical afterlife and Louis XIV by founding Gobelins as a personal luxury-goods production plant. At the other end of the economic scale, the poor often have no choice but to fend for themselves. Wherever you go in the world, you will find inspiring examples of the 'necessity is the mother of invention' principle of design ingenuity. The roads of Asian and African cities are filled with rusty bicycles that have been converted into ersatz trucks and people carriers. Farmers have protected their land for centuries by constructing drystone walls from stones they have found on nearby land. A well-built wall can last for decades, marking boundaries and providing shelter for wildlife, as well as for the insects, mosses and lichen that live among the stones.[9]

In developed economies, customization was the norm for rich and poor alike until the Industrial Revolution, when mechanization made it possible to manufacture huge quantities of standardized objects more cheaply and efficiently than producing them individually. At the time, uniformity seemed seductively new, which is one reason why late eighteenth-century socialites were so besotted by Josiah Wedgwood's factory wares. Standardization felt even more alluring to early twentieth-century modernists, for whom it was no longer a novelty but seen as a means of building a better future. Their optimism was shared by the management theorists of the day, notably the American engineer-turned-author Frederick Winslow Taylor. Having declined a place at Harvard to become an apprentice machinist, he rose up the ranks of a Philadelphia steelworks by spotting clever ways of improving its operations. Taylor distilled what he learnt there in his 1911 book *The Principles of Scientific Management*, which proposed standardizing every aspect of management and production, including the design of the finished products.[10] Among his admirers was a young Detroit automotive engineer Henry Ford, who put Taylor's principles into practice at the Ford Motor Company, where he defined the 'Fordist' formula for running an efficient assembly line by combining standardized design and production – just like Qin's weapon makers – with relatively generous wages for his workers and competitive prices for the cars.[11] 'Any customer can have a car painted any colour that he wants so long as it is black,' said Ford, after discovering that waiting for the paint to dry took up more time than any other part of the manufacturing process, and that black was the fastest-drying shade.[12] Such ploys helped the price of the Model T Ford to fall from just over $800 in 1908 to $250 in 1919, making it affordable for most skilled workers in the United States.

Over half a century after the publication of Taylor's book, the legacy of Fordism could be seen in a 1965 film of Martha and the Vandellas performing their single 'Nowhere to Run' on the Mustang production line at Ford's huge Baton Rouge plant. The manufacturing process continued uninterrupted while Martha Reeves, Rosalind Ashford and Betty Kelly sang the song. Dancing past a row of painters aiming spray guns at panels of the cars, they jumped into a half-built Mustang on a hydraulic assembly line. Clicking their fingers, they sang from the car's seats while the workers put panels into position, fitted wheels and winched in the engine. By the time the song ended, the car was finished and Martha and the Vandellas jumped out, making way for its delighted 'owner'. They waved him off as he drove away in his new Mustang. The song lasted for less than three minutes.[13]

By then, the corporate enthusiasm for standardization was reinforced by the official edicts of national standards institutes, safety regulators and consumer watchdogs, as well as a labyrinth of legislation imposed by individual countries and international bodies such as the European Union to regulate the size, weight, density, power and other facets of countless products and services. The benefits have been immense, not only by giving millions of people access to cheaper products, but by making many of them safer, more efficient and less damaging to the environment.

The downside is that standardization can also make our lives seem soulless. Take the colours of the fabrics used to upholster the chairs designed by Charles and Ray Eames in the post-war era. When Hella Jongerius analysed the colours used by Vitra, which manufactures the Eameses' furniture in Europe, she noted that technically they were the same shades specified by the designers years before, even though they looked subtly different. The hues of the original fabrics were slightly variegated, which made them appear warmer and more nuanced. Vitra's technicians had continued to use the same shades, but they had gradually become more homogeneous as the chemical specifications of the dyes and finishes were tweaked to comply with changes in safety regulations. Jongerius's solution was to combine marginally different versions of the same shades to recreate the original effect.[14]

Compared to improved safety, greater reliability and the other practical advantages offered by standardization, marginally duller furniture is a minor concern. But even tiny issues like this can grate in an age when, having enjoyed the benefits of standardization for so long, we take them for granted, and yearn for idiosyncrasy. The popularity of vintage fashion, the revival of interest in craft and folklore, and the fad for dangling trinkets from mobile phones are all attempts to stamp our personalities on our possessions, as are the crazes for 'hacking' tech products by modifying them and 'steampunking' objects by presenting them in fantastical, antique guises. And professional designers have become adept at finding ways of enabling us to personalize the things that fill our lives on a larger scale by assuaging our desire for individuality, without sacrificing the practical benefits of standardization.

Sometimes, it is enough simply to give us the impression that an object is unique. Hella Jongerius designed the various patterns on her Repeat upholstery fabric for the American textile company Maharam so that none of them would be repeated for long enough to spoil the illusion that each length of cloth is distinctive. She also programmed the production process of her B-Set dinner service to add tiny flaws to the glazing. After decades of industrial 'perfection', those glitches look so incongruous that we associate them with the charming quirks of handcrafted ceramics even though they are identical on each piece.[15] The Japanese retailer Muji achieves a similar effect, by omission, not inclusion. The name Muji is an abbreviation of *mujirushi*, which means 'no brand' in Japanese, and there is no visible branding on its products.[16] Once you have bought something from a Muji store and removed the price sticker, all traces of its origins disappear, and it feels as though that product really is yours.

A similar transition has been made in architecture. In Farshid Moussavi's book *The Function of Form*, she describes how Mies van der Rohe applied Fordist principles to the construction of buildings by using the then-innovative method of steel frame construction.[17] He also experimented with different configurations of prefabricated steel components to open up the interiors to their surroundings: the frenzy of New York's streets in the case of the Seagram Building on Park Avenue, and the natural beauty of the Czech countryside for Villa Tugendhat in Brno.[18] Towards the end of his career Mies was criticized for repeating similar forms, specifically in the Bacardi Building in Bermuda and the Neue Nationalgalerie in Berlin. He replied by stating that he 'refused to design a new architecture every Monday morning'. Whereas Rem Koolhaas's practice OMA conceived the China Central Television headquarters in Beijing so that the building would never seem quite the same. If you look at the structure, it appears to distort slightly each time you change the angle at which you are seeing it, or move closer towards it, or further away. As a result, whenever you see the building it feels intensely personal because you know that no one else, not even you, will ever observe it in quite the same way again. Impossible though it seems for something as massive as the CCTV building to move or morph, the complex geometry of OMA's design fosters the impression that it does just that.[19]

In the 1960s and early 1970s, the avant-garde architecture group Archigram sketched fantastical projects like the Plug-In City, which had a limitless capacity to expand or contract as different units were plugged in and out.[20] The French brothers Ronan and Erwan Bouroullec have applied similar principles to familiar objects by designing modular products that can be personalized by their users when they add or subtract various elements to change their size or purpose. The Joyn office system that the brothers designed for Vitra was inspired by the long wooden table in the kitchen of their grandparents' farm in Brittany. At any one time, several people might have been eating at the table while another prepared food and a couple of children pored over their homework. Joyn consists of a series of desks of differing lengths to which screens can be added, to create work-stations of varying sizes and provide privacy in open-plan offices and studios. The space allocated to each person can be adjusted whenever people join or leave the team, and the desks cleared to create a conference table for meetings.[21]

Other designers have deployed software programs to ensure that the same object can be reproduced in an infinite number of ways. Breeding Tables, a project by the Swedish-German design team led by Reed Kram and Clemens Weisshaar, uses computer-controlled laser-cutting technology to create distinctive bases for each of a series of tables. The actual shape is chosen at random by the software.[22] A similar principle was applied to My Private Sky, a collection of bespoke plates they designed for the German porcelain maker Nymphenburg. My Private Sky is a set of seven plates that collectively depict an astrological map of the sky on the night when the owner was born. The date, time and place of birth are punched into a computer, which draws a digital map of the hundreds of stars, planets and galaxies in the sky that night to be painted on to the plate by one of Nymphenburg's artisans.[23]

An alternative is to introduce a spontaneous element to the development process, as the Slovakian designer Tomáš Gabzdil Libertíny did by delegating the making of his Honeycomb Vase to a hive of forty thousand bees. Having cast solid beeswax into the shape of a conventional vase, he placed it inside a hive, where the bees determined the final shape by adding layer after layer of beeswax on to the mould. Beautiful and unique though each vase is, Libertíny's template will only ever work on a small scale. The vases have to be made in April, May and June, when the bees are at their most active. Even then, finishing each one can take as long as a week.[24]

Personalized objects like these tend to be expensive, but there are more affordable approaches. We customize smartphones and computers instinctively, by loading them up with different software programs, apps, files and screensavers. The physical structure of the device remains identical, but the contents of its screen will be distinctive. Even if you and I downloaded the same apps on to our phones, we would arrange them in different ways, according to the frequency with which we expected to use them and where we thought they looked best. Glimpsing the screen of someone else's phone or computer is akin to reading their diary, or scrutinizing the content of their bookshelves. Why else would lecturers look so embarrassed when their graphic interfaces are accidentally projected on to the screen? Even if the contents are uncontentious, they may prefer not to display them in public: doing so risks revealing more about themselves than they might wish.

Graphic interfaces change constantly, depending on whatever we happen to have parked on our screens at particular times. More than any other aspect of a digital product, it is they that make it feel as if it is 'ours', because we have decided what will be seen on them, not the designer or manufacturer – though no graphic interface will feel quite as personal as one containing an app you have designed yourself, as the tens of thousands of mostly self-taught developers of applications have discovered.

The app phenomenon typifies the do-it-yourself spirit that has already fuelled the crazes for tinkering, steampunking, hacking and Maker Faires, and put pressure on even the largest companies to open up their development processes. Once it was deemed the height of corporate chic for products to be developed in deepest secrecy. Steve Jobs relished whipping up speculation about Apple's product development plans by refusing to disclose details in advance and revealing the finished object to a rapturous audience. Spectacularly effective though it was, corporate mores have changed, and many areas of design have embraced the open source process, whereby the development of a design project is open to public scrutiny for other people to comment on and learn from. Open source design was pioneered in the software industry, partly thanks to the efforts of Dennis Ritchie, who insisted that 'Hello, world' and the other programs in C language should be 'open systems' that could be used on different types of computers. Ben Fry and Casey Reas have applied open source principles to the development of Processing, as has Poonam Bir Kasturi in her work on Daily Dump.

In the same spirit of transparency, some companies have utilized digital technology to enable customers to personalize their products from the outset. Nike has an online service with which online consumers can choose different colours, materials and finishes for sports shoes. If you click on part of a running shoe, such as the heel clip or ankle cuff, a menu flashes up explaining the possibilities. You can change the colours, or pick a translucent sole rather than an opaque one. It is also possible to customize the functional attributes of some shoes, such as soccer boots, whose soles can be given extra cushioning to reduce the pressure applied by the studs, or less cushioning for faster response times.[25] Other businesses have used crowd-sourcing techniques to enable their customers to influence the design process, like Local Motors in Arizona, which invites people to contribute ideas to the development of energy-efficient cars.[26]

The most radical approach to customization is the development of exceptionally fast, precise and flexible manufacturing processes, such as three-dimensional printing. By rendering it as easy and inexpensive to make things individually or in very small quantities as in very large ones, 3D printing promises to reverse the commercial logic of the economies of scale that has made standardization so profitable since Josiah Wedgwood's day. The *Economist* has described it as a transformative technology like 'the steam engine in 1750 – or the printing press in 1450, or the transistor in 1950', and predicted that 3D printing could herald the start of a new era of mass customization.[27]

The first stage of 3D printing is to download the design template for an object on to a computer and to tweak it. Maybe you will decide to change the colour or finesse the shape. Once you have adjusted it to suit your wishes, you press 'print' to instruct a nearby 3D printer to make it, and collect the finished piece from there, just as if you were printing a document from a computer. The location of the printer is chosen with your convenience in mind, not the manufacturer's or designer's, It makes no difference if it is thousands of miles away from them, or in a neighbouring building. The object is built by adding fine layers of material on top of one another, not unlike the way in which the bees constructed the delicate beeswax structure of the Honeycomb Vase for Tomáš Gabzdil Libertíny. A similar process, known as 'rapid prototyping', has long been used to produce highly detailed models of buildings for architects and of cars for automotive design studios. One limitation is that, so far, 3D printing is only suitable for specific types of plastic, resin and metal. Another is that it can make only solid blocks of material, such as an iPhone case, but not its contents. A third constraint is cost. Until recently, 3D printers were too expensive to be used on anything other than an experimental basis, but their prices have fallen significantly and will be even lower in future.

The technology is already being used to customize objects of dramatically different sizes and degrees of complexity. It can personalize simple, everyday products, such as cups and pens, by making stylistic tweaks.[28] Other objects can be adapted to make them easier to use by people with physical impediments. If you have arthritic hands, you are likely to have difficulty gripping cooking utensils like knives or spoons. The process can tackle that problem by widening the handles until they are easier to hold. The extreme precision of 3D printing also enables it to produce larger, more complicated components, including aircraft and car parts, which will be capable of withstanding extremes of temperature, weather, speed, weight and tension. These components are often lighter than existing ones, thereby reducing the fuel consumption of the aeroplane or vehicle. Similarly, the technology can be used to make tiny medical devices which need to fit neatly into particular parts of the body, such as hearing aid cases and dental crowns, as well as artificial hips and jawbones. The curvature of a human bone is impossible to replicate using conventional manufacturing techniques, whereas 3D printing can not only match it exactly, but reproduce the lattice-like internal structure to create an implant which is stronger, lighter, and sits more comfortably inside the body. As a result, it also promises to transform the production of prostheses, like Aimee Mullins and Hugh Herr's artificial legs.

As the technology advances, it will be possible to apply it to a wider range of materials and objects. And as the cost of 3D printers falls, they will appear in more and more places, where many more people will have access to them. Eventually, they could be installed in villages and towns all over the world for the local communities to use.[29] Once, people would go to a nearby blacksmith's forge to have things made or repaired; in future they may do the same at a nearby 3D printer. Similarly, construction projects need no longer be held up if there is a delay in delivering an important part, or a faulty one needs to be replaced, because the specifications can be sent immediately to the closest 3D printer.

The growing use of such processes is likely to have an aesthetic impact on the products that fill our lives. Just as leaps in design and construction software have enabled architects such as OMA, Zaha Hadid, SANAA and Farshid Moussavi to construct buildings in complex forms, 3D printing will develop a new vocabulary of shapes for objects.[30] These three-dimensional versions of the eerily intricate digital forms we see on our computer screens will be the contemporary equivalents of other shapes that have come to dominate design whenever new technologies have enabled designers to produce something different: the smooth geometry of 1920s 'machine age' furniture; the luscious curves of 1960s plastic 'pop' pieces; and the ubiquitous 'blobs' that defined product design in the 1990s when some designers got carried away with their newly acquired design software. The work of OpenStructures, Unfold and other design groups committed to developing new ways of working with new production technologies suggests that the surreal aesthetic of 3D printing will also be adept at reflecting the cultural shift away from the twentieth-century illusion of clarity and uniformity by expressing the contradictions and inconsistencies of human nature,[31] as conceptual designers like Anthony Dunne and Fiona Raby have done in their experimental projects.

Design has already influenced the early development of 3D printing, and its contribution will become increasingly important as the technology evolves. A critical challenge for industrial design is to ensure that manufacturers and consumers are able to make the most of the practical benefits of such processes to develop products that are more efficient, less expensive, more durable and better suited to their users' needs. Impressive though all of that sounds, 3D printing also offers an important opportunity to make progress on the sustainable front.

If the only material used in the manufacturing process is the exact quantity needed to make the object in question, there will be no waste, which is not only beneficial environmentally, but financially too because it reduces the risk of manufacturers wasting money on superfluous raw materials. Similarly, if more products are ordered on a bespoke basis, there will be less need for manufacturers and retailers to carry stock which could end up being scrapped should it remain unsold. Nor will it be necessary for them to ship as many products or components from far-flung subcontractors to warehouses and stores, thereby saving fuel. Each product can also be expected to last for longer because repairs will be simpler. All in all, there should be substantial savings in energy, materials and other resources, with tremendous environmental benefits.

The possibility of making products in tiny quantities also promises to usher in a new era of experimentation, as designers and manufacturers become increasingly confident about testing new or unusual ideas. They will be able to gauge the response to a few pieces, or to variations of a design proposal, before making the hefty investment required to put them into volume production, if they still consider that to be necessary in the dynamic new age of ecologically smart customization.

13 What about 'the other 90%'?

Why do the majority of the world's designers focus all their efforts on developing products and services exclusively for the richest 10% of the world's consumers? That question always reminds me of a quote attributed to the bank robber Willie Sutton when someone asked him why he robbed banks. His answer was: 'Because that's where the money is.'
—Paul Polak[1]

If you asked a classroom of seventeen-year-olds what they had last made themselves, what would you expect them to say? Maybe something straightforward like a cake, a birthday card or, if they were into computing, an app. But when Emily Pilloton put that question to her students on her first day as a teacher at the Bertie Early College High School in Windsor, North Carolina, on 11 August 2010, some of them could not remember ever having made anything. 'They'd never held a hammer or taken an art class,' she recalled. 'Half of them didn't even know how to read a ruler.'[2]

This was bad news for Pilloton, who had just moved from San Francisco with her partner Matthew Miller to teach in Windsor, a run-down town in Bertie County, one of the most economically depressed rural areas of the United States, with a history of racial tension. One in three children in Bertie lived in poverty, and the few available jobs were low-skilled ones in agriculture or biotechnology.[3] Two of the thirteen students in the class had to care for babies or toddlers, including one with a four-year-old daughter. Pilloton and Miller, who were twenty-nine and thirty-three at the time, had gone to Windsor to set up and run an experimental design course called Studio H. They had chosen to base it there because the area's economic and social problems were so grave, but as the course was to be structured around the students' experience of making things, the discovery that many of them lacked basic skills was discouraging.

Pilloton was accustomed to working in difficult conditions. Having entered design through the conventional route after studying architecture at the University of California, Berkeley, and product design at the School of the Art Institute of Chicago, she had decided to focus on social and humanitarian concerns. 'At graduate school, people were starting to talk more about sustainability, but I felt it lacked a human factor. Can we really call $5,000 bamboo coffee tables sustainable? After school, I tried really, really hard to work in design and architecture firms, but I'm a contrarian. I'm not good with authority, and didn't really want to have a boss.'[4] She started to make furniture from recycled materials, then blogged about sustainable and humanitarian design, before deciding to address those issues as a designer.

Starting with $1,000 in savings and a makeshift desk on the dining table in her parents' home in Kentfield, California, she founded Project H Design on 8 January 2008. At first, she envisaged it as a design firm specializing in humanitarian endeavours, but swiftly realized that it could be more useful by galvanizing other designers to work on similar projects. Soon she met the team at Architecture for Humanity, including Miller, who was volunteering there, and its co-founders, Cameron Sinclair and Kate Stohr. They offered to lend her a desk in their San Francisco office. There, she observed how AfH functioned by encouraging architects around the world to join local 'chapters', each of which initiated its own projects as well as contributing to collective ones, such as emergency reconstruction programmes after the South-East Asian tsunami and Hurricane Katrina.[5] Suspecting that there were lots of 'like-minded designers who don't want to design doorknobs all day', Pilloton envisaged Project H as a smaller version of AfH, which would focus on design's potential to help the disadvantaged, rather than architecture's.

The learning curve was steep, not least as they were pioneering new approaches to design in extreme environments. When Project H was asked to redesign the quiet rooms in a Texas foster-care home, its initial ideas were dismissed as unsuitable. 'We went in with music, aromatherapy and all this designeresque stuff,' said Pilloton. 'Then the head of the home said: 'You can't have anything that can be used as a missile, or that a kid can choke on, and you've got to use this material on the walls, because we have to think about flying faeces. It was very humbling.'[6]

Another early assignment was to develop an educational playground for the Kutamba AIDS Orphans School in a remote village in the Rukungiri district of southern Uganda, which had been built by AfH with Miller as project leader.[7] As the school's resources were so sparse, the playground had to lend itself to a range of learning experiences, as well as being cheap and easy to construct, and robust enough to survive with minimal maintenance. Project H's chapter in New York devised a series of maths-based games to be played on the Learning Landscape, a square grid of sixteen disused tyres half buried in a sandpit, which could double as an outdoor classroom, with the students and their teacher perching on the tyres.[8]

The concept proved so successful in Uganda that Project H was asked to install Learning Landscapes elsewhere. 'It's an incredibly dumb construction method, but it's cheap – we built one in the Dominican Republic for $75 – and very effective,' said Pilloton. 'We developed games for kids to play in the grid . . . to help them learn anything from kindergarten maths to eighth-grade algebra.'[9] When Sidney W. Zullinger, superintendent of the Bertie County School District, spotted the Learning Landscape on a blog, he invited Project H to build them in four elementary local schools and to design computer labs for three high schools. Working in the cash-strapped North Carolina education system, and seeing how bleak the students' prospects were, prompted Pilloton and Miller to consider a more ambitious intervention. In September 2009 they presented a proposal to run a design course as part of the high-school curriculum, and developed the idea into Studio H in discussion with Zullinger's team and the students and teachers at the Bertie Early College High School. The students were to design and make boards for cornhole, a popular local game, during the first term, and chicken coops in the second. The grand finale was to be the construction of a farmer's market in the centre of Windsor during the third term.[10]

There was never any suggestion of the course preparing the students for careers in design, not least as their chances of finding suitable jobs locally would be so scarce. Instead, it was intended to nurture skills that could help them with problem solving, communication, leadership and innovation in the hope that their experience of studying design and the design process would give them practical and intellectual tools to draw on for the rest of their lives.

Pilloton and Miller had known from the start that they would have to move to Bertie County to run the course. 'There was no way we could have swept in from the West Coast once in a while. We'd have had no credibility with the community.' But they had a lot to give up. Not financially: they were living on a shoestring, running Project H from their laptops, and borrowing AfH's meeting room when necessary, while living in an old Airstream trailer parked on land owned by a friend at Half Moon Bay, south of San Francisco. The network ran on $46,000 of donations (mostly small ones of $50 or less) in year one, and $86,000 in year two, but it was flourishing, having become involved with a string of projects including a water programme in Africa as well as the Texas care-home work and the Learning Landscapes. Once Studio H began, there was a risk that they would have so little time for Project H that it would lose momentum. Nonetheless, they went ahead.[11]

Disaster loomed when, a few weeks before the course was due to start, Zullinger left his post. Having lost their local champion, they had to plead with the school board to allow them to proceed as planned. Eventually the board agreed, and they raced to convert a disused car-body shop behind the school into a classroom, workshop and design studio, where the students would attend Studio H for three hours of every school day. It opened on time, only for Pilloton and Miller to discover that some of the kids they were counting on to build a farmer's market in a few months' time had no idea what to do with a ruler.

By the end of the first term, ten of the original students had completed the cornhole board assignment, but three others had left the course. Two had to leave the school because of failing grades, and a third was excluded from Studio H for aggressive behaviour. 'We had a strict "No Horseplay" rule and they'd all signed contracts saying: "One incident and we're done,"' recalled Pilloton. 'It broke my heart to see him go, but we had no choice.'[12]

The remaining students tackled the chicken coops in their second term, which required more sophisticated carpentry, welding and metalwork, as well as community research and computer work. Originally the coops were intended for use by everyone in Windsor, but when floods destroyed large tracts of the town they were donated to the worst-affected families. Not that it was the only natural disaster to strike Bertie County during the course of the school year. Pilloton and Miller also helped with the local relief effort after a hurricane and a tornado, while conducting a running battle to rebuff the criticism of Studio H from local people who either did not understand what they were trying to achieve there, or disapproved.

Studio H's first two terms were intended to prepare the students for their final assignment: designing and building the

Windsor Super Market, a two-thousand-square-foot wooden pavilion on an empty plot of land near a cluster of community projects on the riverfront. One of the students had hatched the idea of providing a place where local people could sell home-grown produce as well as cakes, bread, pickles and jams. 'The one thing that there is no shortage of around here is land. It is easy for people to grow produce, but there wasn't anywhere to sell it.'[13] The students had researched what sort of market would work best in Windsor, and identified suitable vendors, who started selling their wares near the site in May to raise awareness of the market before the pavilion was completed the following October.

At the opening ceremony the Mayor of Windsor presented a key to the city to the Studio H team, as a formal thank you.[14] 'We saw the students change from a complete lack of understanding and, in some cases, a complete lack of interest on the first day into amazingly well-rounded creative thinkers and communicators,' said Pilloton. 'They'd each come such a long way and felt much more invested in their local community, having built something permanent they could be proud of.'[15] The boy who had been expelled from the course slipped in towards the end of the ceremony, and later sent a letter of apology to her and Miller.

After the school year ended, one student, who had come close to dropping out of school in tenth grade but stayed on to join Studio H, won a place to study agricultural sciences at North Carolina State University. Other students also went on to college, many of them the first members of their families to do so, or trained to pursue a trade they had discovered during their work at Studio H.[16] Pilloton described it as: 'A really wonderful year, but the hardest year of my life. Constructing a two-thousand-square-foot space was physically very demanding, and Matthew and I had to grow into many different roles: social workers, educators, politicians, accountants and local citizens. We immersed ourselves in a place where the conditions and stresses are extreme, but that's what you have to do if you are serious about humanitarian design.'[17]

After running Studio H for a second year in Bertie County, she and Miller moved back to California to re-establish the course at the REALM Charter School in Berkeley. Having raised all the funding for the first two years of Studio H from charitable foundations and individual donors, they did not feel that it could continue in Bertie County without some form of financial support from the school administration. When it was not forthcoming they moved to a more receptive home.[18]

Emily Pilloton and Matthew Miller are unusually plucky, resolute, ingenious and persuasive. Without those qualities, it would have been impossible for them to have developed Project H so swiftly or to have executed as many programmes in different parts of the world, let alone to have taken on as demanding an endeavour as Studio H. They are not the only designers to have such strengths, but they are in the vanguard of a new generation who are determined to devote their skills to improving the lives of the very people who, up until now, have benefited the least from design.

An ugly truth about the design profession is that, until recently, most designers have devoted most of their time, energy and talent to the wealthy minority of the global population, which arguably needs their help the least. Of the seven billion people in the global population, over six billion cannot afford to buy basic products and services, and half of them lack regular access to food, shelter and clean water. Yet design, which is, in theory, well equipped to help to address such problems, has largely ignored their plight by failing to rise to the challenge of what the social entrepreneur Paul Polak has described as 'designing for the other 90%'.[19]

Not that design has neglected the needy entirely. A cherished goal of the modern movement was to help the disadvantaged, but mostly this ambition was realized in the industrialized world, where hundreds of millions of people have benefited from designers' efforts to develop cheaper, more efficient and sustainable goods. Some designers chafed against this, including those who flocked to the Institute of Contemporary Arts in London on 29 November 1963 to discuss 'First Things First', a manifesto written by the graphic designer Ken Garland, which ended: 'We hope that our society will tire of gimmick merchants, status salesmen and hidden persuaders, and that the prior call on our skills will be for worthwhile purposes.'[20] Yet until recently, it was relatively rare for design projects to focus on empowering 'the other 90%', and rarer still for them to originate in the developing economies where most of those people live.

Emily Pilloton and Matthew Miller's work with Studio H is typical of the new wave of activist design projects that are rooted in local communities, as is Something & Son's experiment at FARM:shop in Dalston. A different approach has been adopted by another group of socially engaged designers, who are deploying design thinking as a strategic tool to develop new approaches to delivering critical public services, or to address political concerns, as Charles Booth once did with his London Poverty Maps, and Otto Neurath and Marie Reidemeister in their work at the Isotype Institute.[21]

In the forefront is Participle, whose founder, the social scientist Hilary Cottam, was introduced to design at university, where she was taught by Robert Chambers, a development specialist who had encouraged people to visualize ideas by sketching them or making models when conducting research in rural India. 'He had this breakthrough idea that if you get people to physically model or draw their ideas, you'll have a different dialogue with them – more honest and less defensive,' she explained. 'Designers use those tools instinctively.'[22] Cottam rediscovered design in the mid 1990s, while working for the World Bank in Africa. At the time, there was a significant investment in building infrastructure and she soon realized that the quality of its design was critical in determining its effectiveness. After returning to Britain in 1997, she noted that standards of design were equally poor in many areas of the public sector there, and started to experiment with ways of using it as a tool to improve public services, including schools and the prison system, not only in terms of the buildings and their contents, but of the way those institutions were run. She developed her thinking while running an experimental programme for the Design Council, and founded Participle in 2007.[23]

Participle begins each project by identifying a problem and conducting what Cottam calls 'quick and dirty research' to assess the situation and how it might be addressed. The results are then pitched to prospective partners and funders, to raise the money required to design and test alternatives, and to prototype them. 'One of the things we have always done is to set our own briefs rather than receiving them from others,' said Cottam. 'Otherwise, design always seems to be given the wrong problem to solve, and to find itself at the wrong end of the food chain.'[24] The projects are run by multidisciplinary teams, which always include business specialists as well as designers and, when necessary, social scientists, psychologists, ethnographers, economists, statisticians, technologists and anthropologists. Typically, each team is led by a designer, and uses design techniques and design terminology throughout the development process.

An early priority for Participle was to tackle the problems

Participle's Circle project to improve
the provision of care for the elderly

associated with the rapidly ageing population in Britain and other developed countries. The work began in Southwark, one of London's poorer boroughs, with the aim of designing better ways of caring for the elderly that would cost no more than the existing service provided by the local council. Participle's research showed that the two most important ingredients for a happy, healthy old age had little to do with money, but were ensuring that each person had a social network of at least six people who they saw regularly, and had no need to worry about niggling problems, such as fixing a leaking tap. 'If you take away the little hindrances that can stop an elderly person with limited energy from doing other things, you can make a huge difference,' said Cottam.[25]

Participle's solution was to create a 'Circle' to which up to a thousand elderly residents would belong. It would combine the roles of a concierge service, social club, cooperative and self-help group. When new recruits joined, they would be visited by an existing member to discuss what sort of help they required, and which skills they could contribute to the group. One person may have lost a partner who had taken care of cooking or the couple's finances, and needed to be taught those skills, but was an experienced gardener who could share their knowledge with fellow members. Another might need someone to sit with them when plumbers or electricians came to their home to do repairs, or to be taught how to keep in touch with distant friends and family on email and Skype. In principle, each member of the Circle should receive the support they need, as well as meeting new people and feeling empowered by helping others.[26]

It sounds like common sense, and seems so simple that it is astonishing no one thought of it before. But what does it have to do with design? Traditionally, the only contribution a designer would have made to such a project would have been to produce a logo or leaflets explaining the concept to prospective participants. But designers were involved with every stage of developing the Circle concept. Their skills and experience proved invaluable in talking to local seniors during the initial research, analysing what sort of services they needed, working out what form they might take and how they could be delivered, and then persuading the other people to back it. First, Participle had to convince Southwark Council, the Department for Work and Pensions and a corporate sponsor to invest £850,000 in developing the concept. It then faced the challenge of persuading local seniors, many of whom were deeply sceptical, to participate in an unfamiliar project which, at that point, had no track record. 'Designers understand people's motives, aspirations and needs,' said Cottam. 'They're also natural lateral thinkers, which is critical in projects like this, and their communication skills help to persuade people to participate. Design is only ever one tool in the mix, but it brings something very special.'[27]

The Circle concept proved so successful in Southwark that Participle has since introduced it to other areas,[28] and developed equally ingenious solutions to other social problems. One project aims to help under-achieving teenagers to be more confident and resourceful. Another attempts to resolve the difficulties of families in chronic crisis. Participle's work has not only made a significant difference to the lives of its participants, and to their local communities, but it acts as resounding proof of design's power to address such complex challenges.

A third approach to designing for 'the other 90%' is for designers from developed economies to work within developing ones, as Pilloton and Miller did at the Kutamba AIDS Orphans School in Uganda, and Nathaniel Corum has done by contributing to AfH's relief and reconstruction programmes. Again, there are historical precedents. Buckminster Fuller always advocated the importance of seeking global solutions to problems, and intended his geodesic domes to provide emergency shelter in developing as well as developed countries. And after Otto Neurath's death, Marie Reidemeister extended the Isotype Institute's work into Africa, notably by undertaking a series of projects in Western Nigeria during the period in the 1950s when it was making the transition from being a British colony to an independent state.[29] The progressive government there was committed to providing free primary education and health care, and to modernizing agriculture. The institute produced a series of Isotype booklets to explain their new rights to the local people and to encourage them to make the most of them.[30]

Within architecture, some of the greatest buildings of the twentieth century were designed for developing economies. Among them are Louis Kahn's interventions in the Indian city of Ahmedabad and in the Bangladeshi capital Dhaka, and the work of Le Corbusier and Pierre Jeanneret in Chandigarh, the model modern city they built on a sweltering plain between the Indian states of Punjab and Haryana. As well as producing an inspiring symbol of the new India[31] and a pleasant, orderly place for people to live, their work there trained a new generation of Indian architects. Tragically, much of the furniture and fittings they designed for use in Chandigarh has disappeared from the city after being bought, often for a pittance, by Western art dealers, who then sold them at inflated prices. A cast-iron slab that once covered a Chandigarh drain sold at an Artcurial auction in Paris for $25,000 in October 2011 – not what Le Corbusier and Jeanneret would have wanted.[32]

For many years, the simplest way for Western designers to work in the developing world was to volunteer for organizations such as the Peace Corps in the United States and Voluntary Service Overseas in Britain, which enabled them to contribute to projects there, though not necessarily by applying their design skills. Even so, specialists from other disciplines, such as anthropologists, ethnographers, ecologists and economists, were becoming aware of design's value to economic development. Among them was Paul Polak, who, after a career in psychiatry, founded International Development Enterprises in 1982 as a non-profit organization, which was committed to helping people to move out of poverty by financing new ventures to develop farming equipment.[33] Design swiftly became a core concern when he realized, just as Hilary Cottam had done at the World Bank, how important it was to ensure that the equipment was of the right quality.

By the early twenty-first century, a new network of specialist organizations had emerged with a commitment to help designers to empower 'the other 90%'. Volunteer networks like AfH have enabled thousands of architects – and designers too – to work on disaster relief and longer term development projects on a full-time or part-time basis.[34] Students have contributed to similar programmes as part of their studies through initiatives such as Studio 804 at the University of Kansas, where a group of graduate students completes a social housing scheme each year.[35] New funding sources – such as the Acumen Fund, which provides loans and investment to ventures intended to address global poverty[36] – created more opportunities for designers to participate in existing social enterprises and to initiate their own projects.

Awareness of the field was heightened by the publication in 2006 of a series of books exploring the potential – and pitfalls – of humanitarian design, including John Thackara's *In the Bubble*, Alex Steffen's *Worldchanging: A User's Guide for the 21st Century* and Kate Stohr and Cameron Sinclair's *Design Like You Give a Damn*.[37] During the same year, Paul Polak coined the concept of 'designing for the other 90%', while addressing the

Aspen Design Summit. 'I said that 90% of the world's designers spent all their time addressing the needs of the richest 10% of the world's customers,' he recalled. 'I also said that before I die I want to see that silly ratio turned on its head.'[38] The Smithsonian Cooper-Hewitt National Design Museum in New York adopted his phrase as the title of an exhibition, 'Design for the other 90%', which opened the following year.[39]

By far the boldest humanitarian design project was taking shape at this time. One Laptop per Child was launched as a new non-profit organization devoted to the development of an educational computer, which was to be sold to governments all over the world for less than $100. It was the brainchild of Nicholas Negroponte, the American technologist who co-founded the MIT Media Lab, and who unveiled the plans for OLPC at the World Economic Forum's annual meeting in the Swiss village of Davos in January 2005.[40] Inspired by the work of his MIT colleague Seymour Papert, Negroponte's objective was to distribute laptops to the world's poorest children on the principle that a computer would be the single most useful tool to help them to learn. Not only would they acquire basic computing skills, but they would have access to the research resources of the Internet and a digital library of electronic books. During the early 1980s, Negroponte and Papert had travelled to Senegal, taking a batch of Apple II computers to be given to children there, and had studied their responses.[41] A year after his initial announcement, Negroponte returned to Davos with a mock-up of the proposed laptop to sign a memorandum of understanding with the United Nations Development Program. He also claimed to be close to clinching agreement with Argentina, Brazil, China, Egypt, India, Nigeria and Thailand to order some seven million computers.[42]

Negroponte was adamant that the quality of the laptop's design would determine whether or not OLPC was successful. Functionally, the design team had to develop a laptop that could be made cheaply enough to be sold for as little as $100, while providing all of the learning resources that children needed and being tough enough to survive the extreme weather and the rough and tumble of a child's life in a developing country. The laptop had to be as attractive as possible, to look at and to use, so that each child would want to take care of it. 'We want our laptops to be as cool as the iPod,' he said. 'Nothing less.'[43]

As well as cajoling his contacts into providing funding for the project, Negroponte assembled a network of suppliers, subcontractors and designers to collaborate with the OLPC team. Lisa Strausfeld, who had studied under Muriel Cooper at MIT (and named her daughter after her), worked on the software design; and the Swiss product designer Yves Béhar, whose company, fuseproject, was based in San Francisco, worked on the case and other aspects of the hardware. The development process was conducted on an open source basis so that observers could comment on it.[44]

Expectations were high when the first XO-1 laptops left their Shanghai production line in November 2006, but they lived up to Negroponte's promises. Cutely cartoonish in style, they were encased in bright green and white plastic with pop-up antennae that resembled a rabbit's ears. Each component served several functions. The antennae doubled as USB port covers and latches to close the case; and the bumper both as a seal and a place for kids to rest their hands while scrolling. The laptop itself could also be used as an electronic book, a television set and a light, which, OLPC hoped, would encourage families to hold on to the machines long after their children had left school.

One of the biggest design challenges was minimizing manufacturing costs. The design team made the XO-1 smaller than a conventional laptop and added flash memory, but no hard drive, leaving it with less storage capacity. Another cost-cutting measure was to reduce the processing power. Even so, OLPC needed to secure substantial orders to achieve the economies of scale required for the XO-1 to hit its target price. The orders for seven million laptops announced by Negroponte at Davos had not materialized, but OLPC hoped to secure orders for five million, the minimum number required to start production, and to sell the laptops for $150 each. It then planned to obtain enough additional orders to cut the price to $100 by 2008.[45]

What was there not to love? OLPC was dazzlingly ambitious, and clearly determined to deliver what it had promised. Its cause was magnificent: helping the world's poorest children to fulfil their educational potential in the hope of transforming their lives and those of their families, eventually steering their countries towards economic growth and political stability. And design was at the heart of the endeavour. What a wonderful way of demonstrating its potential to help 'the other 90%'.

OLPC won shoals of design awards for the XO-1, and swamped the design media and conference circuit. When the first laptops were delivered to their owners, various glitches were discovered, mostly minor ones. The plastic feet had an irritating tendency to slide off desks. The keyboard was too flimsy, and the case too slippery. And kids had difficulty identifying their machines because they all looked the same. The design team found suitable solutions: replacing the plastic feet with rubber ones; strengthening the keyboard with a steel plate; adding a 'goose bump' surface to the case to make it easier to grip; and devising four hundred different colour combinations for the cases to differentiate them.[46] A similar list of problems and improvements was applied to the software design.

Feted though it was in design circles, OLPC had been beset by criticism since the start. The tech industry accused it of distorting competition with its low prices, while claiming that it would be impossible for OLPC or anyone else to make a laptop for as little as $100. Environmentalists warned of the danger of children discarding broken or unwanted laptops unless suitable arrangements were made to repair them, and to dispose of them safely. The development camp accused OLPC of imposing First World solutions on the Third World. And educationalists questioned whether it was justifiable to expect the governments of poor countries to spend what would be a hefty percentage of their education budgets on computers, when so many schools lacked other resources. They also asked if a computer really was the most effective learning tool for the students that OLPC wished to help. Why not spend the same money on buying books, or on hiring more teachers?

Some of the criticism was self-serving, such as the tech industry's grumbles about unfair competition, or fuelled by resentment among other non-profit groups that Negroponte and OLPC had succeeded in raising so much money so swiftly, possibly at their expense. But other complaints were valid, and OLPC tried to address the functional flaws, for example by training local repair and maintenance teams to service the laptops. But it soon became apparent that its biggest problem would be securing enough orders to hit the $100 price target.

One difficulty was that some of the promised 'orders' had vanished. Changes of government, economic squeezes and old-fashioned *braggadocio* were to blame. Even if all the orders had been completed, OLPC may have had to raise the price above $150, because of factors beyond its control such as the impact of the declining US dollar on the cost of labour, raw materials and components. The XO-1 began full-scale production in November 2007. Six months later, OLPC had delivered laptops to three hundred thousand children, rather than the expected millions, with each one costing $187.[47]

OLPC ploughed on, cutting its overheads while developing new models and devising smart marketing ploys – such as a 'Give One Get One' programme whereby people in the developed world could buy a laptop at a premium price which also paid for one to be donated to a child in a developing country.[48] There were positive reports from teachers whose students were using XO-1s, and from the children themselves. Yves Béhar remembered receiving 'incredible letters from teachers telling us how the laptops changed their method of teaching' and 'letters from kids who say that they won't let go of their laptops under any circumstances'.[49] Teachers reported that rather than having to lecture continuously throughout a class, they could speak for, say, ten minutes, then leave the kids to check references on their XO-1s for five minutes, which made the lesson more compelling and reduced discipline problems. Unexpectedly, the XO-1 also won over its critics in the tech industry, when it was credited as being a catalyst for the surge in sales of tablets, notebooks and other smaller computers – though this was not what OLPC had planned or hoped for.

By November 2009, a million laptops had been delivered. Nicholas Negroponte claimed that his initial sales targets were 'knowingly hyperbolic'. Had he been less ambitious, he said, the project would have had less impact and found it harder to attract government attention.[50] Even so, OLPC was locked in a dispiriting cycle. The easiest way to increase orders was to reduce the price, but it could not do that unless it had more orders.

One ray of hope was the first substantial order for some four hundred thousand XO-1s for all the primary school children in Uruguay as part of a government programme to ensure that every school student and teacher in the country had a laptop. There were some logistical difficulties. The first fifty thousand machines were equipped with English rather than Spanish software, and connectivity was so patchy, especially in rural areas, that many students had difficulty going online. In some schools, only half of the laptops could access the Internet at the same time. Others had to bus their students elsewhere for exams. Teachers also complained that they had not received adequate training. Even so, the Uruguayan government was confident enough to order more laptops for secondary school students. OLPC could finally analyse the impact of its computers at scale in a long-term project, and assess what support was required in terms of training, repairs, maintenance and learning resources.[51]

By early 2012, OLPC had shipped over two million laptops to students in forty-five countries, but the price of each one had risen to as much as $229. It was also completing development of a radically different model, the XO-3, a tablet computer that it was hoping to sell for $100. By then, a group of students at the Indian Institute of Technology in Jaipur had received the first batch of another tablet, the Aakash, which the Indian government had purchased from its manufacturer, Datawind, for roughly $44 each.[52] India was among the countries that OLPC had initially expected to place a substantial order only to develop a similar project of their own.

The saga of OLPC has been both a boon and a bane for humanitarian design. A boon because its audacity, coupled with its pedigree and engaging design, have acted as lightning rods in attracting the attention of the media, the public and donors not only to its cause, but to others too. Would the Aakash have been developed without it? Or any of the other inexpensive tech products designed for use in developing countries? Possibly not, at least not as many, and not as swiftly. But OLPC has been a bane for the obvious reason that it has failed to match its own expectations. Whether or not Negroponte really was being 'knowingly hyperbolic' when setting such lofty targets, he doomed OLPC to fail, or to look as though it had, even though putting laptops in the hands of more than two million children, most of whom would not otherwise have had access to computers, would be considered a remarkable achievement by any other measure.

Not that OLPC has been the only high-profile humanitarian design project to have struggled. The Aakash has too. Even worse was the plight of PlayPump, which was developed in South Africa during the early 1990s as a means of providing clean water to communities that needed it. Water was pumped out of the ground by a borehole pump driven by a children's roundabout. Whenever kids played on the roundabout, it span around and water was pumped up into an elevated tank. The walls of the tank doubled as advertising billboards, whose rental income was intended to cover the cost of maintenance.[53] It was an engaging idea, illustrated by fetching images of children providing badly needed water while playing happily. In 2006, the US President's Emergency Plan for Aids Relief announced a plan to raise $60 million to install PlayPumps across Africa.[54]

The problem was that extracting water from a PlayPump required a lot of children to push a roundabout for a long time, far longer than the kids in most communities were willing or able to devote to it. Some places had overestimated the amount of water that a single PlayPump could provide, and ended up with too little. The advertising revenue was lower than expected, and the maintenance costs higher. The charity Water Aid issued a critical report on PlayPump, and negative media articles appeared. PlayPumps International, the organization behind the system, closed in 2010 and gave its inventory to the charity Water for People.[55]

To a far greater degree than OLPC, the PlayPump debacle reinforced the post-colonial stereotype of do-gooding Westerners pouring sorely needed resources into well-intentioned but misguided projects in developing countries that did not meet the needs and wishes of the people who were intended to use them, or the nuances of their context. Such accusations are familiar to anyone who has worked in economic development, and, as the humanitarian design field expanded, a fiery debate erupted about the merits and demerits of different approaches.

The US design commentator Bruce Nussbaum, who had been a Peace Corps volunteer in his youth, raised important questions in a 2010 post for Fast Company's Co.Design blog, entitled 'Is Humanitarian Design the New Imperialism? Does our desire to help do more harm than good?' 'Are designers the new anthropologists or missionaries, come to poke into village life, "understand" it and make it better – their "modern" way?' he wrote. 'Should we take a moment now that the movement is gathering speed to ask whether or not American and European designers are collaborating with the right partners, learning from the best local people, and being as sensitive as they might to the colonial legacies of the countries they want to do good in?'[56] All excellent points, which have been fiercely debated for many years within different areas of the development community.

By the 2010s, tens of thousands of designers and design students had engaged in projects intended to empower 'the other 90%', and many design schools had introduced courses on humanitarian and social design. For young designers, volunteering for such ventures had become a standard element of gap years and holidays as well as of their coursework. Some designers, like Nathaniel Corum,[57] had chosen to devote their working lives to humanitarian causes, but the flexible nature of networks like AfH and Project H enabled others to participate on a part-time basis or during sabbaticals from their full-time jobs. Commercial design groups, including IDEO and fuseproject,[58] set aside time and resources for their staff to contribute to pro bono projects.

Critically, a growing number of gifted and resourceful designers have emerged in developing economies, determined to use their skills to improve the quality of life for their compatriots. Traditionally, the design culture of those countries has been dominated by resourceful individuals, often working anonymously and in isolation. Their work was largely unsung, and rarely made available to enough people to achieve its full potential. Many projects suffered from a dearth of investment, which made it difficult to test ideas thoroughly and to improve them when necessary. And such endeavours are rarely mentioned in design books, most of which hail from the developed world. Even Victor Papanek, who was so visionary in other aspects of ethical design, admitted in the preface to the 1985 second edition of his best-selling book *Design for the Real World* that 'much of what I wrote about design for the Third World in this book's first edition now seems somewhat naïve'.[59] Specifically, he felt that he had failed to appreciate the value of what he and fellow Westerners could learn from their peers in developing countries about identifying design solutions to problems both there and in developed economies.

The new generation of designers in developing countries are proving Papanek's point by planning and executing increasingly smart and ambitious projects. India is in the forefront, partly because of the historical strength of its design schools, such as the Indian Institute of Technology in Jaipur and the National Institute of Design in Ahmedabad, where Poonam Bir Kasturi studied; and partly because of the long entrepreneurial tradition there, which fostered her work on Daily Dump and encouraged her to establish it as a catalyst for other enterprises as well as a model of design and design thinking.

Similar principles have driven a Ugandan venture, Eco-fuel Africa, which was founded by the social entrepreneur Sanga Moses to produce clean, cheap cooking fuel and organic fertilizer for poverty-stricken rural regions like the one where he was born. Growing up in one of Uganda's poorest villages, Sanga did not wear shoes until he was thirteen, and was fifteen when he first watched television. Two years later in 1999, he started his first business with just $200. The first member of his family to go to university, he studied business administration and set up several ventures after graduating, including a rural computer school. He also enrolled on an organic charcoal-making course, which gave him the knowledge and practical skills required to launch Eco-fuel Africa.[60]

The crux of the project is to encourage farmers to collect agricultural waste, which is converted into organic fuel and fertilizer in equipment designed by Eco-fuel Africa. The farmers are given inexpensive, easily operable kilns to carbonize their waste into organic charcoal, some of which they keep to fertilize their land, with the rest being sold to Eco-fuel Africa. The kilns are provided free, with the farmers undertaking to pay for them from the proceeds of their charcoal sales. Eco-fuel Africa converts the charcoal into fuel in the form of briquettes, using energy-efficient compressors. The briquettes are then sold to local people at affordable prices to be used as cooking fuel.[61]

The system was carefully planned to address problems that have long haunted rural communities in sub-Saharan Africa. Firstly, farmers can increase their income by selling the charcoal and by improving the productivity of their soil with the fertilizer. Secondly, new jobs are created to deliver charcoal for the farmers as well as to operate Eco-fuel Africa's compressors and distribute the briquettes. Thirdly, many more people have access to inexpensive cooking fuel, which is cleaner and safer than the dirty, smoky scraps of wood and dung they usually burn. Every year, over one and a half million Africans die from indoor air pollution, often having been poisoned by the noxious fumes of makeshift cooking fuel. Fourthly, fewer people will have to forage for kindling. Traditionally, women and children in rural areas of sub-Saharan countries like Uganda have devoted large chunks of their days to doing so, often in potentially dangerous areas.[62] A few years ago, Sanga returned to his village to visit his mother and saw his twelve-year-old half-sister carrying a bundle of firewood on her head. She had walked over six miles to a nearby town to buy it, and the same distance back, which meant that she had missed school that day. If her family no longer needed her to collect wood, she could use her time more productively, to study and, eventually, to earn money.[63] Finally, by reducing the need to remove trees and scrub, Eco-fuel Africa's work should help to tackle the problem of deforestation in Africa, which Sanga also plans to address by planting new trees.

Having successfully established the system in one region, Sanga plans to introduce it throughout Uganda, and then in the rest of East Africa. His goal is to provide more than forty million Africans with clean, affordable energy, though he shares Bir Kasturi's ambition to foster entrepreneurialism too. Eco-fuel

Africa has prototyped a franchising concept in a village by giving the residents a compressor and teaching them how to produce enough fuel to meet their energy needs.[64] For years, visionary projects like this one were stymied by the dearth of funding in developing countries, but Sanga Moses and his colleagues have overcome those constraints by applying design and design thinking with great aplomb, as Daily Dump has done in India. Like Studio H's community programme in the United States and Participle's social design projects in Britain, their work demonstrates how the intelligent use of design can improve the lives and future prospects for 'the other 90%': the people who need it the most.

Epilogue
Redesigning design

I believe that design is problem solving with grace and foresight. I believe that there is always a better way. I believe that design is a human instinct, that people are inherently optimistic, that every man is a designer, and that every problem can either be defined as a design problem or solved with a design solution.
— Emily Pilloton[1]

Forgetting to take your pills; mistakenly taking too many, or too few; taking them at the wrong time of day; forgetting to check the expiry date; fumbling with a new inhaler because you have forgotten how you were told to operate it: for one reason or another, and mostly because of forgetfulness, one in two prescription medicines are taken incorrectly, with potentially devastating consequences for our health.

Mathieu Lehanneur was so struck by that statistic as a design student in Paris that he decided to address the underlying problems in his graduation project. But when he told the director of his design school what he was planning, the response was discouraging. The director advised him to pick another theme. Why would the pharmaceutical industry be interested in a designer's thoughts on encouraging patients to take their medicines correctly? Or the medical profession, come to that? If Lehanneur wanted to apply design to health care, he should stick to packaging and brand identities.[2]

Doubtless, the advice was well intentioned, but Lehanneur ignored it and developed a series of devices to help people to take the correct doses at the right time. Some were as simple as printing the prescription on the bag in which the drugs were carried, and dividing a course of medicine into a string of plastic 'beads' with the relevant date printed on each daily dose. Another solution was to infuse a 'therapeutic' paper handkerchief with hay-fever medicine so that sufferers could alleviate their symptoms each time they used it.[3] Far from being ignored by the pharmaceutical industry, a decade later the project was being adapted for production, and Lehanneur was advising a Paris hospital on ways of improving its care of terminally ill patients. The hospital has since introduced LED screens to a palliative care unit showing what the sky will look like the following day. Even patients who may not live that long feel as though they have a stake in the future by dint of knowing something about it, and the friends and relatives who come to visit them always have something to say in what might otherwise be a distressing encounter.[4]

By demonstrating design's agility at developing unexpected yet effective solutions to complex problems, Mathieu Lehanneur makes a compelling case for designers to be given greater responsibility in new areas of our lives. So do Hilary Cottam and the Participle team by reinventing critical areas of social services; Ben Fry and Casey Reas by helping to tackle the data deluge crisis; Emily Pilloton and Matthew Miller by empowering the disadvantaged; Poonam Bir Kasturi and Sanga Moses by devising ingenious solutions to environmental problems; and all the other enterprising designers who have written their own job descriptions as auteurs, activists, conceptualists, adventurers, theorists, strategists, social reformers, ecologists, mavericks and entrepreneurs.

The curator Paola Antonelli has predicted that in years to come designers, like physicists, will fall into one of two categories: theoretical and applied. Both camps will act as what she calls 'pragmatic intellectuals' by analysing important social, political and technological issues that could affect our quality of life, such as food crises and dwindling natural resources, but the former will do so abstractly, and the latter by planning and executing practical solutions to the problems.[5] Such a scenario seems closer thanks to the achievements of Lehanneur, Cottam and their peers.

There will be no shortage of challenges for designers to wrestle with. Dwindling natural resources. Freak weather. Abuses of digital privacy. Data deluge. Crumbling social services. Bloated landfill sites. Paralysed roads. Congested airports. Computer viruses. Technophobia. Economic imbalances. Fissile communities. Food shortages. Imperilled species. So much space junk hurtling around outer space that the International Space Station has had to change its axis to avoid collisions. Designers will be tussling with these and other problems for decades to come.

They will have powerful tools to help them in the scientific discoveries, which will flow from the underwater drones now scouring the deep seas, the Blue Brain Project in Lausanne, the Large Hadron Collider, the International Space Station and research laboratories all over the world. In its 'Hello World' role, design can help to translate those leaps in scientific knowledge into things that could make our lives more efficient, enjoyable and responsible. It can do the same by ensuring that future developments in robotics, nanotechnology, supramolecular chemistry, biomimicry and digital production technologies are put to constructive use, as past generations of designers did for the transistor and computer.

In essence, design will remain the same: still an agent of change, as it was in Qin Shihuangdi's day; and still, as a 1593 reference in the *Oxford English Dictionary* put it: 'a plan or scheme conceived in the mind of something to be done'.[6] But if it is to realize its true potential, it needs to evolve. The old-fashioned virtues of integrity, efficacy, ingenuity and appropriateness will be as important as ever, if not more so, but other qualities must be nurtured too.

Among them is openness. In the open source era, secrecy no longer looks like an alluring exercise of power, but a symptom of insecurity and cause for suspicion. Designers need to be less defensive, and more amenable to collaboration with other disciplines, not as tactical ploys to move into new terrain, but as learning opportunities. The same generosity should be extended to the 'accidental' designers, who have practised design by chance, and, critically, to the people who use design and, thanks to the new wave of manufacturing technologies, will exercise greater influence over its outcome as they experiment with customization.

Design must also become more compassionate. Old-school design was defined by certainties, as you would expect of a culture that was fired by modernist fervour and intent on improving the lives of millions of people by dint of standardization. At its best, this culture was plucky and optimistic, but it also erred towards arrogance, obduracy and boosterism. Those qualities will prove even more damaging in future. Design needs to become more empathetic, and better attuned to the frailties that defy rational analysis yet determine so many elements of our lives, such as making half of us prone to muddling up something as simple and important as taking prescription medicine correctly.

And design needs to be both bolder and humbler. Designers should never be censured for ambition or courage, both of which will be required in abundance if they are to aim higher in terms of the scale and intensity of their work. Yet they must also exercise restraint in accepting design's limitations. If not, its credibility in its new spheres of influence will disintegrate. Nor should designers shirk the mundane aspects of noble endeavours, such as helping to build a sustainable society.

Design must also become more diplomatic. Designers can be excellent communicators, as they have proved over the centuries in the symbolic coup of the skull and crossbones, and all the maps, signage systems and visual languages that have guided, enlightened and protected us. Yet they have been conspicuously less successful at championing their own cause; hence the muddles and misleading clichés that have dogged design over the years. Taking on bigger, more onerous challenges will raise the stakes, because the consequences of failure will be so much graver. Designers need not only to exercise greater rigour in their work, but to be more adroit when disseminating it.

Daunting though these obstacles are, none of them is insurmountable. And it is in all of our interests that design succeeds in overcoming them, and in fulfilling its potential to become a wiser, more constructive force in our lives.

Eco-fuel Africa's system of producing clean, safe organic cooking fuel

Author's note

It always strikes me as odd that as powerful a force as design should so often be misunderstood and trivialized. Design determines the quality of so many aspects of our lives, and as we cannot avoid its influence, the more fully we understand it, the likelier we are to be able to turn it to our advantage.

The more I have learnt about design, the more intrigued I have become, particularly in the seven years that I have had the pleasure of writing a weekly column about design for the *International Herald Tribune*. Many of the themes I have discussed there are addressed in this book, but in greater depth than a newspaper column allows.

Anyone who writes about design has to explain what they mean by it. In *Hello World*, as in the *International Herald Tribune*, I have interpreted it as broadly as possible, not only as a professional discipline, but also as an intuitive process that existed long before there was a word to describe it. This book addresses design's role in planning third-century BC military campaigns and helping eighteenth-century pirates to terrorize their victims into speedy surrender, as well as in developing new objects, information systems, and smarter ways of dealing with data deluge and recycling trash. Although architecture is an important area of design, it is addressed only fleetingly because it operates on such a different scale to other fields that I felt it might be distorting to include it.

Discovering the power and nuances of design has given me great happiness over the years. I hope that reading *Hello World* will help others to share my enjoyment.

Acknowledgements

It would have been impossible to write *Hello World* without the help and encouragement of many people, above all Simon Prosser, who has been a wonderfully empathic and inspiring publisher. I am immensely grateful to Simon and his colleagues at Hamish Hamilton and Penguin – in particular to Anna Kelly, Caroline Craig, Caroline Pretty and Marissa Chen – as well as to my literary agent Derek Johns and his team at AP Watt.

One of the pleasures of publishing this book has been the chance to work with one of my favourite designers, Irma Boom. I was delighted when she agreed to design *Hello World*, and am thrilled with the result. I am deeply grateful to Irma not only for designing the book so beautifully, but for being such fun to work with. I also owe special thanks to my dear friend Paola Antonelli for being so generous with her time and knowledge in reading and commenting on an early draft of the text.

I would like to thank my colleagues at the *International Herald Tribune* and *New York Times* Media Group, and all of the friends who have supported me throughout the process of researching and writing *Hello World*, with special thanks to the following for their contributions: Stuart Comer; Hilary Cottam; Emily King and Matthew Slotover; Michael Maharam and Sabine Steinmair; Farshid Moussavi; Aimee Mullins; Louise Neri; Hans Ulrich Obrist; and Emily Pilloton.

My thanks are due to the staff of the institutions where I conducted much of the research for the book: the Hyman Kreitman Reading Rooms of Tate Library at Tate Britain in London; the Museum Library of the Museum of Modern Art in New York; and the National Art Library at the Victoria & Albert Museum in London.

I am immensely grateful to all of the individuals and institutions who have so generously allowed us to use their images in *Hello World*: Matthew Barney and Gladstone Gallery; Black Mountain College Museum + Art Center and the Estate of Hazel Larsen Archer; Christoph Büchel and Hauser & Wirth; Mariana Cook; Ben Fry; Jill Greenberg; IBM Corporate Archives; Juliet Kepes Stone and the György Kepes Foundation; Mathieu Lehanneur; Tomáš Gabzdil Libertíny; The Library of the London School of Economics & Political Science; Julia Lohmann; Christien Meindertsma; Aimee Mullins; Hattula Moholy-Nagy and the Estate of László Moholy-Nagy; Sanga Moses and Eco-fuel Africa; the Otto and Marie Neurath Isotype Collection, Department of Typography & Graphic Communication, University of Reading; Participle Ltd; Project H Design; Something & Son; Stamen; Thonet Gmbh; Unfold; University of Manchester; and Z33.

Many remarkable designers, historians and commentators have inspired my thinking and writing on design, as have their colleagues in other fields. I owe them all a huge debt, especially: Ken Adam; David Adjaye; Ai Weiwei; Gail Anderson; Lisa Armstrong; Edward Barber; Paul Barnes; David Batchelor; Yves Béhar; Berg; Michael Bierut; Iwona Blazwick; Adam Bly; Achim Borchardt-Hume; Ronan and Erwan Bouroullec; Constantin and Laurene Leon Boym; Michael Braungart; Tim Brown; Eliza Brownjohn; Buckminster Fuller Institute; Margaret Calvert; Emily Campbell; Matthew Carter; Aric Chen; David and Evelyn Chipperfield; Nathaniel Corum; Michael Craig-Martin; Ilse Crawford; Eames Demetrios; Farrokh Derakhshani; Chris Dercon; Digital Forming; Anthony Dunne; Ben Evans; Helen Evenden; Rolf Fehlbaum; Jamie Fobert; Studio Formafantasma; Elena and Norman Foster; Christopher Frayling; Tobias Frere-Jones; Ben Fry; Naoto Fukasawa; Tomáš Gabzdil Libertíny; Beatrice Galilee; Martino Gamper; Piero Gandini; Nicolas Ghesquière; Graphic Thought Facility; Konstantin Grcic; Joseph Grima; Joost Grootens; Martí Guixé; Steve Haake; Pierre Hardy; Jaime Hayon; Thomas Heatherwick; Hugh Herr; Jonathan Hoefler; Richard Hollis; Gary Hustwit; Kigge Hvid; INDEX: Design to Improve Life; David James; Hella Jongerius; Frith Kerr; Eric Kindel; Kram/Weisshaar; Clémence and Didier Krzentowski; Joris Laarman; Mathieu Lehanneur; Amanda Levete; Armand Limnander; Liu Zhizhi; Emma Loades; Julia Lohmann; William McDonough; Brendan McGetrick; John Maeda; Enzo Mari; Bruce Mau; J. Mays; Christien Meindertsma; Alessandro Mendini; Andy Merritt; Metahaven; M/M (Paris); Bill Moggridge; Jasper Morrison; Murray Moss and Franklin Getchell; Nicholas Negroponte; Marc Newson; Elaine O'Hanrahan; Jay Osgerby; Oscar Peña; James Peto; Julia Peyton-Jones; Phoebe Philo; Paul Polak; Potion; Fiona Raby; Dieter Rams; Carlo Ratti; David de Rothschild; Lindy Roy; Zoë Ryan; Sanga Moses; Peter Saville and Anna Blessmann; Paula Scher; Louise Schouwenberg; Libby Sellers; Tom Shakespeare; Britte Siepenkothen; Raf Simons; Cameron Sinclair; Paul Smith;

Paul Smyth; Lisa Strausfeld; Studio Museum Achille Castiglioni; Stefano Tonchi; Edward Tufte; Suneet Singh Tuli; Alexander von Vegesack; Wedgwood Museum; Alasdhair Willis; and those who wish to remain anonymous. Thank you.

Notes

Prologue Hello World

1. Martin Pawley, *Buckminster Fuller: How Much Does the Building Weigh?* (1990; London: Trefoil Publications, 1995), p. 12.
2. Elizabeth Flock, 'Dennis Ritchie, Father of C Programming Language and Unix, Dies at 70', *Washington Post*, 13 October 2011, http://www.washingtonpost.com/blogs/blogpost/post/dennis-ritchie-father-of-c-programming-language-and-unix-dies-at-70/2011/10/13/gIQADGNbhL_blog.html; Steve Lohr, 'Dennis Ritchie, Trailblazer in Digital Era, Dies at 70', *The New York Times*, 13 October 2011, http://www.nytimes.com/2011/10/14/technology/dennis-ritchie-programming-trailblazer-dies-at-70.html

1 What is design?

1. László Moholy-Nagy, *Vision in Motion* (Chicago, Ill.: Paul Theobald, 1947), p. 42.
2. Robin D. S. Yates, 'The Rise of Qin and the Military Conquest of the Warring States', in Jane Portal (ed.), *The First Emperor: China's Terracotta Army* (London: British Museum Press, 2007), p. 31.
3. *Ibid.* pp. 42–50.
4. Michael Loewe, 'The First Emperor and the Qin Empire', in Portal, *The First Emperor*, pp. 75–8.
5. Hiromi Kinoshita, 'Qin Palaces and Architecture', in Portal, *The First Emperor*, pp. 84–5.
6. Jessica Rawson, 'The First Emperor's Tomb: The Afterlife Universe', in Portal, *The First Emperor*, pp. 118–29.
7. Lukas Nickel, 'The Terracotta Army', in Portal, *The First Emperor*, pp. 158–79.
8. Henry Dreyfuss, *Designing for People* (New York: Simon & Schuster, 1955), p. 14.
9. William Little, H. W. Fowler and Jessie Coulson, *The Shorter Oxford English Dictionary on Historical Principles*, vol. 1, ed. C. T. Onions (Oxford: Clarendon Press, 1987), p. 528.
10. John Heskett, *Toothpicks & Logos: Design in Everyday Life* (Oxford: Oxford University Press, 2002), p. 5.
11. The phrase 'intelligent design' had been bandied about by scientists for years in various guises, but in 1989 it appeared in *Of Pandas and People: The Central Question of Biological Origins*, a textbook for US high-school students written by Percival David and Dean H. Kenyon and published by the Foundation for Thought and Ethics in Texas. In the book, they mount a neo-creationist argument that parts of the universe were created, not by the organic process of evolution described by Charles Darwin, but by a rational process orchestrated by an 'intelligent designer'. Most neo-creationists believe that the 'intelligent designer' was the Christian God. Some also contend that conventional interpretations of science were too narrow, and should be broadened to include super-natural phenomena. Their critics have accused them of being ultra-conservative cranks.
12. Giorgio Vasari, *Lives of the Artists: Volume I* (1550; London: Penguin Books, 1987), pp. 256–7. A fulling machine is used in the process of making woollen cloth. It cleans the cloth by scouring it repeatedly to remove oil, dirt and other impurities before thickening it.
13. Heskett, *Toothpicks & Logos*, p. 5.
14. Timo de Rijk, *Norm = Form: On Standardisation and Design* (Den Haag: Foundation Design den Haag; Gemeentemuseum Den Haag; Uitgeverij Thieme Art b.v., Deventer, 2010), pp. 32–5.
15. Ian Thompson, *The Sun King's Garden: Louis XIV, André Le Nôtre and the Creation of the Gardens of Versailles* (London: Bloomsbury, 2006), p. 125.
16. Peter Burke, *The Fabrication of Louis XIV* (New Haven, Conn.: Yale University Press, 1992), pp. 51–8.
17. Alison Kelly (ed.), *The Story of Wedgwood* (1962; London: Faber and Faber, 1975), p. 18.

18. Margarita Tupitsyn, ‘Being-in-Production: The Constructivist Code’, in Margarita Tupitsyn (ed.), *Rodchenko & Popova: Defining Constructivism* (London: Tate Publishing, 2009), p. 13.
19. Penny Sparke, *Italian Design: 1870 to the Present* (London: Thames & Hudson, 1988), pp. 10–14.
20. ‘Icons of Progress’ (2011), IBM, http://www.ibm.com/ibm100/us/en/icons/gooddesign/
21. Reyner Banham, *Design by Choice*, ed. Penny Sparke (London: Academy Editions, 1981).
22. Roland Barthes, *Mythologies* (1957; Frogmore, St Albans: Paladin, 1973); Jean Baudrillard, *The System of Objects* (1968; London: Verso, 2005).
23. Robert Venturi, Denise Scott Brown and Steven Izenour, *Learning from Las Vegas: The Forgotten Symbolism of Architectural Form* (1972; Cambridge, Mass.: The MIT Press, 1977).
24. Janet Abrams, ‘Muriel Cooper: Biography by Janet Abrams’, AIGA website to mark the award of the 1994 AIGA Medal to Cooper, http://www.aiga.org/medalist-murielcooper/; reprinted from ‘Flashback: Muriel Cooper’s Visible Wisdom’, *I.D. Magazine* (September–October 1994).
25. Interview with John Maeda, August 2007, in Alice Rawsthorn, ‘Muriel Cooper: The Unsung Heroine of On-screen Style’, *International Herald Tribune*, 1 October 2007.
26. ‘Everything I Know’, Buckminster Fuller Institute, 6 March 2010, http://bfi.org/about-bucky/resources/everything-i-know
27. Calvin Tomkins, ‘In the Outlaw Area’, *New Yorker*, 8 January 1966, in K. Michael Hays and Dana Miller (eds), *Buckminster Fuller: Starting with the Universe* (New York: Whitney Museum of American Art, 2008), p.180.
28. Martin Pawley, *Buckminster Fuller: How Much Does the Building Weigh?* (1990; London: Trefoil Publications, 1995), p. 122.
29. Calvin Tomkins, ‘In the Outlaw Area’, in Hays and Miller, *Buckminster Fuller*, pp. 198–9.
30. Mark Oliver, ‘“Slowly but steadily, madness descended”’, *Guardian*, 10 February 2005, http://www.guardian.co.uk/business/2005/feb/10/money.uknews; Sharon LaFraniere, ‘All iPhone Sales Suspended at Apple Stores in China’, *The New York Times*, 13 January 2012, http://www.nytimes.com/2012/01/14/technology/apple-suspends-iphone-4s-sales-in-mainland-china-stores.html
31. Little, Fowler and Coulson, *Shorter Oxford English Dictionary on Historical Principles*, p. 528.
32. Interview with Nathaniel Corum, July 2010, in Alice Rawsthorn, ‘A Font of Ideas from a “Nomadic” Humanitarian Architect’, *International Herald Tribune*, 2 August 2010, http://www.nytimes.com/2010/08/02/arts/design/02iht-design2.html
33. Participle, http://www.participle.net/
34. For example, the American Society of Civil Engineers publishes an annual ‘Report Card’, which assesses the condition of fifteen aspects of public infrastructure in the United States, including drinking water, energy, hazardous waste, roads, schools and transit. The 2009 version produced four Cs and eleven Ds, and calculated that some $2.2 trillion of investment would be needed in the US over the next five years to patch up the damage and make the infrastructure fit for purpose.
35. Tim Brown, *Change by Design: How Design Thinking Transforms Organizations and Inspires Innovation* (New York: HarperCollins, 2009), pp. 6–7.
36. Jenny Uglow, *The Lunar Men: The Friends Who Made the Future, 1730–1810* (London: Faber and Faber, 2002), p. 112.
37. IDEO, http://www.ideo.com
38. Interview with Peter Saville, February 2012.
39. Bernard Rudofsky, *Architecture Without Architects: A Short Introduction to Non-Pedigreed Architecture* (1964; Albuquerque: University of New Mexico Press, 1987); The Museum of Modern Art, Press Archive, 10 December 1964, https://www.moma.org/docs/press_archives/3348/releases/MOMA_1964_0135_1964-12-10_87.pdf?2010
40. ‘Gwangju Design Biennale 2011’, http://gb.or.kr/eng/gdb/. Ai Weiwei initiated the ‘Un-Named Design’ exhibition in his role as joint artistic director of the Gwangju Design

Biennale in South Korea only to be arrested by the Chinese authorities in April 2011 and imprisoned for an indefinite period of time. A group of his collaborators, led by the curator Brendan McGetrick, continued work on the exhibition during his imprisonment. When Ai was released from prison in June 2011, he resumed work on the project the following day. However, the Chinese government rejected repeated requests from the biennale organizers to allow him to travel to Gwangju to supervise the installation of the exhibition that summer and to attend the official opening of the biennale in September 2011.

41. Steve Jones, *Darwin's Island: The Galapagos in the Garden of England* (London: Little, Brown, 2009), pp. 146–8.
42. 'A Short History of the Dalmatian' from the American Kennel Club's *The Complete Dog Book* (1992), posted on the Dalmatian Club of America website, http://www.thedca.org/dal_hist_by_akc.html; Ron Punter, 'Origin and History of the Lakeland Terrier', Lakeland Terrier Club, http://lakelandterrierclub.org.uk/profile.htm; 'Who was Parson Jack Russell?', 'History of the Breed' section of the Jack Russell Terrier website, http://www.jack-russell-terrier.co.uk/breed/who_was_jack_russell.html
43. Benoit Denizet-Lewis, 'Can the Bulldog Be Saved?', *The New York Times Magazine*, 22 November 2011, http://www.nytimes.com/2011/11/27/magazine/can-the-bulldog-be-saved.html
44. Jones, *Darwin's Island*, pp. 81–3.
45. Denizet-Lewis, 'Can the Bulldog Be Saved?'

2 What is a designer?

1. Victor Papanek, *Design for the Real World: Human Ecology and Social Change* (1971; Chicago, Ill.: Academy Chicago Publishers, 1985), p. 3.
2. Together with Keith Richards, Blackbeard is said to have been a role model for the styling of Johnny Depp's character Captain Jack Sparrow in the *Pirates of the Caribbean* movies.
3. Captain Charles Johnson, *A General History of the Robberies and Murders of the Most Notorious Pyrates* (1724; London: Conway Maritime Press, 2002).
4. Interview with Tom Wareham, curator of maritime history at the Museum of London Docklands, April 2011.
5. Jamaica Rose and Michael MacLeod, *A Book of Pirates: A Guide to Plundering, Pillaging and Other Pursuits* (Layton, Utah: Gibbs M. Smith, 2010), p. 137.
6. Interview with Tom Wareham, April 2011.
7. BP Global, http://www.bp.com/; Prada, http://www.prada.com/
8. Jeremy Lewis, *The Life and Times of Allen Lane* (London: Penguin Books, 2006), pp. 244–5.
9. Interview with Rolf Fehlbaum, February 2010, in Alice Rawsthorn, 'A Dash of Color at Vitra's Eclectic Site', *International Herald Tribune*, 15 February 2010, http://www.nytimes.com/2010/02/15/arts/15iht-design15.html
10. Architects can be equally dictatorial. Take Frank Lloyd Wright (supposedly the model for Howard Roark, the megalomaniacal architect played by Gary Cooper in King Vidor's 1949 movie version of Ayn Rand's novel *The Fountainhead*). His cousin, Richard Lloyd Jones, called to complain that the roof of Westhope, a house Wright had designed for him in Tulsa, Oklahoma, in the 1930s, was faulty. 'Dammit, Frank, it is leaking on my desk,' wailed Lloyd Jones. To which Wright replied, 'Richard, why don't you move your desk?' Meryle Secrest, *Frank Lloyd Wright: A Biography* (New York: Alfred A. Knopf, 1992), p. 372.
11. The British novelist Daniel Defoe published a heavily romanticized account of Henry Avery's career in piracy in his 1719 novel *The King of Pirates* (London: Hesperus Classics, 2002). The dashing Douglas Fairbanks played the title role of the 1926 silent movie *The Black Pirate*, as did Errol Flynn in the 1935 film *Captain Blood*. Johnny Depp starred as Captain Jack Sparrow in the *Pirates of the Caribbean* movie franchise from 2003 onwards.
12. John Cooper, *The Queen's Agent: Francis Walsingham at the Court of Elizabeth I* (London: Faber and Faber, 2011), pp. 144–5. Hindlip Hall was the home of the

staunchly Catholic Abington family, and is now the headquarters of the West Mercia Police.

13. Nicholas Owen was canonized by Pope Paul VI in 1970 as one of the Forty Martyrs of England and Wales, all of whom were executed between 1535 and 1679. He was canonized for his courage, selflessness and for saving the lives of so many fellow Roman Catholics.
14. Robert Grudin, *Design and Truth* (New Haven, Conn.: Yale University Press, 2010), p. 107.
15. Jefferson's love of tinkering led to a fascination with design and manufacturing, which influenced his political work. As a young diplomat in Paris during the early 1780s, he had visited a musket workshop, owned and run by a gunsmith called Le Blanc, who had developed a progressive method of ensuring that his muskets were made from identical, interchangeable components. Le Blanc clashed with the French authorities after his rivals goaded them into questioning his methods, but Jefferson's report on his workshop had a lasting impact on manufacturing in the United States. Henry Dreyfuss, *Designing for People* (New York: Simon & Schuster, 1955), p. 21; John Heskett, *Industrial Design* (London: Thames & Hudson, 1980), p. 50.
16. Charles Darwin's family home, Down House, was in the village of Downe in Kent. The Post Office added an 'e' to the village's name, but Darwin refused to do the same. Steve Jones, *Darwin's Island: The Galapagos in the Garden of England* (London: Little, Brown, 2009), p. 2.
17. Jonathan Olivares, *A Taxonomy of the Office Chair* (London: Phaidon, 2011), p. 17.
18. Peter Burke, *The Fabrication of Louis XIV* (New Haven, Conn.: Yale University Press, 1992), p. 58.
19. The Meissen manufactory was founded in 1710 by Augustus the Strong, Elector Prince of Saxony, at Albrechtsburg Castle in Germany. Augustus wished to emulate the finely crafted porcelain which was imported to Europe from China, and the artisans working for his court, including the goldsmith Johann Jakob Irminger, were instructed to design vessels and figures for the manufactory. Other pieces were designed by anonymous craftsmen who were employed there. 'Our tradition', Meissen, http://www.meissen.com/en/about-meissen®/our-tradition
20. By the time Wedgwood was apprenticed to him, Thomas Whieldon was already experimenting with new ways of organizing production at his factory by allotting different workers to specific tasks, such as throwing, turning, handling, modelling and decorating the pots. Similar systems had been used in the early 1700s by the Chinese porcelain industry, and were gradually becoming popular in Europe. The Scottish economist David Hume had analysed the benefits of what he called the 'partition of employments' in his 1739 book *A Treatise of Human Nature*, and his compatriot Adam Smith reaffirmed the importance of the 'division of labour' in *An Inquiry into the Nature and Causes of the Wealth of Nations*, published in 1776. When Wedgwood rented a small factory in 1758 to open his own pottery, he introduced many of Whieldon's experimental ideas there.
21. Jenny Uglow, *The Lunar Men: The Friends Who Made the Future, 1730-1810* (London: Faber and Faber, 2002), pp. 49–52; 'William Hackwood', Wedgwood Museum, http://www.wedgwoodmuseum.org.uk/learning/discovery_packs/___/pack/2182/chapter/2415
22. Uglow, *The Lunar Men*, pp. 349–54.
23. Alison Kelly (ed.), *The Story of Wedgwood* (1962; London: Faber & Faber, 1975), p. 18.
24. 'Frog Service', Victoria & Albert Museum, http://collections.vam.ac.uk/item/O8065/plate-frog-service/
25. Kelly, *The Story of Wedgwood*, p. 34.
26. The most talented modellers honed their skills by working with Wedgwood on replicas of finely crafted antiquities. Their most challenging assignment was to make a copy of the Barberini Vase, an ancient Roman cameo glass vase, which was lent to Wedgwood by its owner the Duke of Portland in 1786. For four years they struggled to reproduce it accurately, developing special materials and production techniques to do so. The Portland Vase, as it was called as a 'thank you' to the duke, is still considered to be among Wedgwood's finest works. Wedgwood held his modellers in high regard,

Hackwood especially. 'Hackwood is of the greatest value and consequence in finishing fine, small work,' he wrote in 1774. Two years later, he wished that he had 'half a dozen more Hackwoods'. 'William Hackwood', Wedgwood Museum.

27. Uglow, *The Lunar Men*, pp. 325–6.
28. Mary Shelley portrayed Victor Frankenstein, the hero of her 1818 novel, as an idealistic medical student whose scientific experiments produce a Creature who bears some physical resemblance to a (grotesque) human being. The novel then explores whether the Creature also shares the human capacity for language, emotions, empathy, vulnerability, the need for companionship, a moral conscience and a soul. Frankenstein sets up a second laboratory in the Orkney Islands where he intends to build a female companion for the Creature, but changes his mind halfway through the project and destroys her. The Creature is so enraged by grief that he kills Frankenstein's friend Clerval and his wife Elizabeth in acts of vengeance. He then flees to the North Pole with Frankenstein in pursuit. The latter dies there, and the Creature disappears after plunging into the ice, never to be seen again. Mary Shelley, *Frankenstein: Or, the Modern Prometheus* (1818; London: Penguin Classics, 2003).
29. Richard Holmes, *The Age of Wonder: How the Romantic Generation Discovered the Beauty and Terror of Science* (London: HarperPress, 2008), pp. 334–5. There were five theatrical versions of *Frankenstein* on the London stage during the 1820s alone.
30. George Eliot, *Middlemarch* (1874; London: Penguin Classics, 1985).
31. The fictional city of 'Milton' was modelled on Manchester, where Elizabeth Gaskell lived for much of her adult life with her husband, William, and their children. The model for 'Darkshire' is the nearby county of Lancashire. Elizabeth Gaskell, *North and South* (1855; Harmondsworth: Penguin Classics, 1987), p. 96.
32. Heskett, *Industrial Design*, p. 19.
33. *Ibid*., pp. 183–4.
34. Martin P. Levy, 'Manufacturers at the World's Fairs: The Model of 1851', in Jason T. Busch and Catherine L. Futter, *Inventing the Modern World: Decorative Arts at the World's Fairs, 1851–1939* (New York: Skira Rizzoli International, 2012), pp. 34–49.
35. 'The Great Exhibition', British Library, http://www.bl.uk/learning/histcitizen/victorians/exhibition/greatexhibition.html
36. 'National Art Library Great Exhibition collection', Victoria & Albert Museum, http://www.vam.ac.uk/content/articles/n/national-art-library-great-exhibition-collection/
37. Richard Sennett, *The Craftsman* (London: Allen Lane, 2008), p. 112.
38. Fiona MacCarthy, *The Last Pre-Raphaelite: Edward Burne-Jones and the Victorian Imagination* (London: Faber and Faber, 2011), pp. 128–32.
39. 'About Dresser', Victoria & Albert Museum's microsite for the 2004 exhibition 'Christopher Dresser: A Design Revolution', http://www.vam.ac.uk/vastatic/microsites/1324_dresser/whoisdresser.html. Christopher Dresser was noted for the quality of his metalware, and particularly for the pieces he developed for companies such as Hukin & Heath and Elkingtons in Birmingham and James Dixon in Sheffield.
40. Achim Borchardt-Hume (ed.), *Albers and Moholy-Nagy: From the Bauhaus to the New World* (London: Tate Publishing, 2006), pp. 163–7.
41. Sibyl Moholy-Nagy, *Moholy-Nagy: Experiment in Totality* (New York: Harper & Brothers, 1950), pp. 64–7.
42. László Moholy-Nagy, *Vision in Motion* (Chicago, Ill.: Paul Theobald, 1947), p. 42.
43. György Kepes's books on visual theory included *Language of Vision*, published in 1944 (New York: Dover Publications, 1995), and a series of essays published in six volumes under the title 'Vision + Value', from 1965 to 1966.
44. Jennifer Bass and Pat Kirkham, *Saul Bass: A Life in Film & Design* (London: Laurence King, 2011); György Kepes (ed.), *György Kepes: The MIT Years, 1945–1977* (Cambridge, Mass.: The MIT Press, 1978).
45. *Time*, 31 October 1949.
46. Raymond Loewy, *Industrial Design* (London: Faber and Faber, 1979), p. 25.
47. Adèle Cygelman, *Palm Springs Modern: Houses in the California Desert* (New York: Rizzoli International, 1999), pp. 38–49.

48. Raymond Loewy, *Industrial Design* (London: Faber & Faber, 1979), p. 36.
49. Olivier Boissière, *Starck®* (Cologne: Benedikt Taschen, 1991), pp. 84–7.
50. Charlotte and Peter Fiell, *1000 Chairs* (Cologne: Taschen, 2000), p. 570.
51. 'Louis Ghost', Starck, http://www.starck.com/en/design/categories/furniture/chairs.html#louis_ghost
52. Victor Papanek pointed out disapprovingly in *Design for the Real World* (p. ix) that Raymond Loewy and other prominent American industrial designers of his day, including Henry Dreyfuss, Harold Van Doren and Norman Bel Geddes, had all started their careers either by designing stage sets for the theatre or department-store window displays.
53. Raymond Loewy, *Industrial Design*, p. 8.
54. Interview with Nathaniel Corum, July 2010, in Alice Rawsthorn, 'A Font of Ideas from a "Nomadic" Humanitarian Architect', *International Herald Tribune*, 2 August 2010, http://www.nytimes.com/2010/08/02/arts/design/02iht-design2.html; for Hilary Cottam, see Participle, http://www.participle.net/
55. Interview with Emily Pilloton, October 2011, in Alice Rawsthorn, 'Humanitarian Design Project Aims to Build a Sense of Community', *International Herald Tribune*, 24 October 2011, http://www.nytimes.com/2011/10/24/arts/24iht-design24.html
56. Hilary Cottam set up Participle in 2007 with Hugo Manassei, whose background is in digital technology.
57. Interview with Hilary Cottam, January 2012.
58. An example is the work of the American designers Ben Fry and Casey Reas, who developed Processing, a new computer programming language that provides a solution to the data crisis by distilling digital information into the luscious, constantly changing digital imagery of data visualizations. Rather than relying on a tech company to distribute it, as Dennis Ritchie and his colleagues did with the computer programming languages they devised at Bell Labs, Fry and Reas released it themselves through their website, http://www.processing.org/
59. The Swiss design theorist François Burkhardt observed that for as long as designers were restricted to commercial roles, their ability to pursue their own would be limited. 'They are very ill-placed to set themselves up as a pressure group able to produce real change,' he wrote, 'for they are no more than employees, working for companies within which they have little autonomy.' François Burkhardt, 'Design and "Avantpostmodernism"', in John Thackara (ed.), *Design After Modernism: Beyond the Object* (London: Thames & Hudson, 1988), p. 147.
60. Tim Brown, *Change by Design: How Design Thinking Transforms Organizations and Inspires Innovation* (New York: HarperCollins, 2009), p. 13.
61. An example is William Joyce, who worked for the animation studio Pixar before co-founding Moonbot Studios in Shreveport, Louisiana, which produced several acclaimed interactive books including *The Fantastic Flying Books of Morris Lessmore* and *The Numberlys*; http://www.moonbotstudios.com/. 'Moonbot Studios Launches in Louisiana', uploaded on to YouTube by louisianaeconomicdev on 12 April 2011, http://www.youtube.com/watch?v=0ueBGHtOkyQ
62. Michael Braungart and William McDonough, *Cradle to Cradle: Re-making the Way We Make Things* (2002; London: Jonathan Cape, 2008).
63. Lilly Reich's achievements were obscured by her close collaboration with her lover, Mies van der Rohe. When he emigrated to the United States in 1938, she stayed behind in Berlin during the Second World War only for her studio to be destroyed in a bomb raid. Two years after the war ended, Reich died at the age of sixty-one. Mies lived until 1969, by which time Reich was largely forgotten, and many of her achievements perceived as being his.
64. 'America Meets Charles and Ray Eames', from NBC's *Home* show, presented by Arlene Francis and broadcast in 1956; uploaded on to YouTube by hermanmiller on 23 November 2011, http://www.youtube.com/watch?v=IBLMoMhIAfM
65. Esther da Costa Meyer, 'Simulated Domesticities: Perriand before Le Corbusier', in Mary McLeod (ed.), *Charlotte Perriand: An Art of Living* (New York: Harry N. Abrams, 2003), pp. 36–7.

66. Megan Gambino, 'Interview with Charles Harrison', *Smithsonian Magazine*, 17 December 2008, http://www.smithsonianmag.com/arts-culture/Interview-With-Intelligent-Designer-Charles-Harrison.html
67. In the United States, more than half the students at most design schools are now female. There are still fewer women at the top of the design profession than men there and in other countries, but the number of influential women designers is growing globally, notably Hella Jongerius in product design, Ilse Crawford in interiors, Paula Scher in graphics and Lisa Strausfeld in software. There has also been progress in terms of ethnic diversity, especially among African American designers. Stephen Burks is a leading figure in furniture design, as is Gail Anderson in graphics and Joshua Darden in typography. But comparatively few black teenagers choose to study design in North America or Europe, possibly because they are deterred by the dearth of role models, instead deciding to pursue careers in fields where they can realistically expect to have a better chance of success. Encouragingly, dynamic designers are emerging in Africa and Asia, including Sanga Moses in Uganda and Poonam Bir Kasturi in India, both of whom have executed sustainable design projects that promise to alleviate poverty and foster entrepreneurialism.

    Tellingly, one area of design where women have thrived is the relatively recent one of software. As well as Muriel Cooper, the pioneers of digital imagery in the late twentieth century included Lillian Schwartz, Vera Molnar and Barbara Nessim. Elsewhere, Hilary Cottam and Emily Pilloton are among the leaders in social design, while Neri Oxman and Daisy Ginsberg are at the forefront of redefining the relationship between design and science.
68. Dennis Ritchie was instrumental in the design of the Unix computer operating system, which he developed at Bell Labs with a colleague, Ken Thompson. Unix later inspired the open source development of the free operating system Linux. Steve Lohr, 'Dennis Ritchie, Trailblazer in Digital Era, Dies at 70', *The New York Times*, 13 October 2011, http://www.nytimes.com/2011/10/14/technology/dennis-ritchie-programming-trailblazer-dies-at-70.html
69. Among the exhibits at Maker Faire Africa in Cairo were an automated car park, energy-efficient traffic lights and a human-powered vehicle that looked not unlike Buckminster Fuller's beloved Dymaxion Car; http://makerfaireafrica.com/about/
70. Walter Isaacson, *Steve Jobs* (London: Little, Brown, 2011), p. 501; 'Apple's Mac App Store Downloads Top 100 Million', Apple, http://www.apple.com/pr/library/2011/12/12Apples-Mac-App-Store-Downloads-Top-100-Million.html
71. 'The Third Industrial Revolution', *Economist*, 21 April 2012, http://www.economist.com/node/21553017

3 What is good design?

1. Reyner Banham, 'H.M. Fashion House', *New Statesman*, 27 January 1961.
2. Fiona MacCarthy, *William Morris* (1994; London: Faber and Faber, 1995), p. 418.
3. William Morris, *Hopes and Fears for Art* (London: Longmans, Green and Co., 1919).
4. Edgar Kaufmann Jnr, *Good Design* (New York: Museum of Modern Art, 1950). The Good Design Awards were founded by Kaufmann Jnr, who had curated the original 'Good Design' exhibitions at the Museum of Modern Art, New York, together with Charles and Ray Eames, Eero Saarinen, George Nelson and other designers.
5. Plato, *Early Socratic Dialogues* (1987; London: Penguin Classics, 2005).
6. C. J. Chivers, *The Gun: The AK-47 and the Evolution of War* (London: Allen Lane, 2010), p. 9.
7. *Ibid.*, pp. 143–200; Max Hastings, 'The Most Influential Weapon of Our Time', *New York Review of Books*, 10 February 2011, http://www.nybooks.com/articles/archives/2011/feb/10/most-influential-weapon-our-time/
8. Chivers, *The Gun*, pp. 313 and 335.
9. Donald Judd, 'It's Hard to Find a Good Lamp' (1993), http://www.juddfoundation.org/furniture/essay.htm. This essay was originally published in the exhibition catalogue *Donald Judd Furniture* (Rotterdam: Museum Boijmans van Beuningen, 1993).
10. 'Doodle History', Doodle 4 Google, http://www.google.com/doodle4google/history.html

11. 'Doodles', Google, http://www.google.com/doodles/finder/2012/All%20doodles
12. Jennifer Bass and Pat Kirkham, *Saul Bass: A Life in Film & Design* (London: Laurence King, 2011), pp. 9–11.
13. 'Saul Bass Title Sequence – The Man with the Golden Arm (1955)', uploaded on to YouTube by Movie Titles on 1 September 2010, http://www.youtube.com/watch?v=sS76whmt5Yc. *The Man with the Golden Arm* was directed by Otto Preminger and produced by Otto Preminger Films, Carlyle Productions.
14. 'Vertigo Start Titles', uploaded on to YouTube by VISUALPLUS 1 on 4 August 2010, http://www.youtube.com/watch?v=5qtDCZP4WrQ. *Vertigo* was directed by Alfred Hitchcock and produced by Paramount Pictures, Alfred Hitchcock Productions (uncredited).
15. 'Casino (1995) opening title', uploaded on to YouTube by reklamtuning on 20 September 2010, http://www.youtube.com/watch?v=HMva00IO0zA. *Casino* was directed by Martin Scorsese and produced by Universal Pictures, Syalis DA, Légende Entreprises, De Fina-Cappa.
16. American Red Cross, http://www.redcross.org/
17. Kaufmann Jnr, *Good Design*.
18. There are two conflicting explanations of why that shade of yellow was chosen for the original Post-it Note. One is that it was the colour of a scrap of paper in the office at the headquarters of 3M, the Post-it's manufacturer, when the colour was being discussed. Another explanation is that 3M expected the principal purchasers of the product to be law firms, and decided to make the original Post-its in the same shade as US legal pads.
19. London 2012 Olympics, http://www.london2012.com/
20. The early version of the London 2012 logo was also criticized on functional grounds. Several people with epilepsy reported suffering seizures after seeing the original animated version on television after the logo's launch in 2007. 'Epilepsy Fears over 2012 Footage', BBC News, 5 June 2007, http://news.bbc.co.uk/1/hi/6724245.stm
21. 'Alternative London 2012 Olympic Games Logo Video', uploaded on to YouTube by Artealee on 12 June 2007, http://www.youtube.com/watch?v=EJSsRILZpRg
22. Helvetica was designed by Max Miedinger and Eduard Hoffmann for the Haas type foundry in Münchenstein, Switzerland. Launched in 1957, it has become one of the most influential typefaces of modern times. Arial was introduced in 1982 by Monotype Typography. Gallingly for design purists, many more people use Arial than Helvetica on their computers, because until recently it was more widely available than the older font, even though the quality of its design is poorer.
23. Mark Simonson, 'How to Spot Arial', Mark Simonson Studio (February 2001), http://www.ms-studio.com/articlesarialsid.html
24. Mark Simonson, 'The Scourge of Arial', Mark Simonson Studio (February 2001) http://www.ms-studio.com/articles.html
25. *In Search of Wabi Sabi with Marcel Theroux*, Part 3, BBC4, first broadcast 16 March 2009, http://www.bbc.co.uk/programmes/b00kvr8m
26. Interview with Konstantin Grcic, August 2006, in Alice Rawsthorn, 'Utility Man', *The New York Times Style Magazine*, 8 October 2006, http://www.nytimes.com/2006/10/08/style/tmagazine/08tutility.html
27. Louise Schouwenberg (ed.), *Hella Jongerius: Misfit* (London, New York: Phaidon, 2010).
28. Naoto Fukasawa and Jasper Morrison, *Super Normal: Sensations of the Ordinary* (Baden, Switzerland: Lars Müller Publishers, 2007).
29. 'Inventor of the Week Archive: Art Fry & Spencer Silver, Post-it® notes', MIT School of Engineering, http://web.mit.edu/invent/iow/frysilver.html
30. Interview with Matthew Carter, June 2006, in Alice Rawsthorn, 'Quirky Serifs Aside, Georgia Fonts Win on Web', *International Herald Tribune*, 10 July 2006, http://www.nytimes.com/2006/07/09/style/09iht-dlede10.2150992.html
31. Robert Grudin, *Design and Truth* (New Haven, Conn.: Yale University Press, 2010), p. 8.
32. Chivers, *The Gun*, pp. 25–6.
33. Raffi Khatchadourian, 'The Gulf War: Were There Any Heroes in the BP Oil Disaster?', *New Yorker*, 14 March 2008, pp. 36–59.
34. Flos, http://www.flos.com/int-en-Home

35. Grudin, *Design and Truth*, pp. 15–17.
36. Arnd Friedrichs and Kerstin Finger (eds), *The Infamous Chair: 220ºC Virus Monobloc* (Berlin: Gestalten, 2010).
37. The biodegradable packaging of the bananas was cited as an example of a design project which tries, but fails, to be sustainable by Cameron Sinclair, co-founder of Architecture for Humanity, at the World Economic Forum's annual meeting in Davos, Switzerland, in January 2010 during a debate on 'Design for Sustainability'. Alice Rawsthorn, 'Debating Sustainability', *International Herald Tribune*, 31 January 2010, http://www.nytimes.com/2010/02/01/arts/01iht-design1.html
38. The Toyota Prius was introduced in Japan in 1997, and worldwide in 2001. 'History of Toyota Prius', uploaded on to YouTube by Nunofos on 30 October 2009, http://www.youtube.com/watch?v=NCtGshT0OpA
39. The Series 7 chair was designed by Arne Jacobsen in 1955, and the Egg in 1958. Charlotte and Peter Fiell, *1000 Chairs* (Cologne: Taschen, 2000), pp. 345–6.
40. Jumana Farouky and Julian Isherwood, 'A Seating Problem at McDonald's', *Time*, 11 October 2007, http://www.time.com/time/business/article/0,8599,1670431,00.html

4 Why good design matters

1. Bruce Mau and the Institute Without Boundaries, *Massive Change* (London: Phaidon, 2004), p. 6.
2. Brigg Reilley, Michel Van Herp, Dan Sermand and Nicoletta Dentico, 'SARS and Carlo Urbani', *New England Journal of Medicine*, 15 May 2003, http://www.nejm.org/doi/full/10.1056/NEJMp030080; Elisabeth Rosenthal, 'The Sars Epidemic: The Path; From China's Provinces, a Crafty Germ Breaks Out', *The New York Times*, 27 April 2003, http://www.nytimes.com/2003/04/27/world/the-sars-epidemic-the-path-from-china-s-provinces-a-crafty-germ-breaks-out.html
3. John Heskett, *Toothpicks & Logos: Design in Everyday Life* (Oxford: Oxford University Press, 2002), p. 9.
4. Aimee Mullins, http://www.aimeemullins.com/index.php
5. Interview with Aimee Mullins, March 2012.
6. The 'toes' in those prostheses were described by Aimee Mullins as 'figurative suggestions formed into the foot shape, not actual digits'.
7. Interview with Aimee Mullins, March 2012.
8. When she described seeing the mannequin at Madame Tussaud's, Aimee Mullins said, 'The important point here is that it was the *calibre* of work done on a Tussaud's mannequin – with the layering of texture to propose tendons and muscles, and the specificity of colour and shape to produce the desired aesthetic – that was not present in prosthetic work where human beings would hugely benefit from such attention to detail. Most "mannequins" are basic human-like forms which only suggest "a leg" without an achilles tendon or toe digits or a calf muscle . . . which is actually what most standard-issue prosthetics have always mimicked, and still do.'
9. Interview with Aimee Mullins, March 2012.
10. 'Aimee Mullins Returns to Dorset Orthopaedic', http://www.dorset-ortho.com/aimee-mullins-returns-to-dorset-orthopaedic/
11. Interview with Aimee Mullins, March 2012.
12. Aimee Mullins modelled in Alexander McQueen's fashion show for his autumn/winter 1999 women's ready-to-wear collection in London.
13. Matthew Barney's film *Cremaster 3* was released in 2002: http://www.cremaster.net/crem3.htm
14. Interview with Aimee Mullins, March 2012.
15. 'Hugh Herr', MIT Media Lab, http://www.media.mit.edu/people/hherr; MIT Media Laboratory Press Archive, 'Powered Ankle-Foot Prosthesis', http://www.media.mit.edu/press/ankle/
16. Interview with Aimee Mullins, March 2012.
17. Graham Pullin, *Design Meets Disability* (Cambridge, Mass.: The MIT Press, 2009), p. 33. The South African athlete Oscar Pistorius, who, like Aimee Mullins, was born without fibulas in both lower legs, faced similar criticism, but succeeded in proving that

in his case his prostheses do not give him an unfair advantage. Pistorius was banned from entering 400-metre events against runners with intact biological legs when he was wearing bespoke J-shaped carbon fibre Flex-Foot Cheetah prosthetic lower legs, but successfully overturned the ban. He made history at the London 2012 Olympic Games by becoming the first amputee athlete to compete in the Olympics. His prostheses have earned him the nicknames 'Blade Runner' and 'the fastest man on no legs'. Michael Sokolove, 'The Fast Life of Oscar Pistorius', *The New York Times Magazine*, 18 January 2012, http://www.nytimes.com/2012/01/22/magazine/oscar-pistorius.html

18. Interview with David James, June 2006, in Alice Rawsthorn, 'A Quest for Perfection for the Most Basic Thing: A Ball', *International Herald Tribune*, 26 June 2006, http://www.nytimes.com/2006/06/25/style/25iht-dlede26.2045652.html
19. The Questra was reportedly 5 per cent faster in flight at the 1994 World Cup than its predecessor had been at the 1990 tournament. Andy Coghlan, 'World Cup Players Face a Whole New Ball Game', *New Scientist*, 9 July 1994.
20. Roger Cohen, 'Germany Opens World Cup with Goals Galore', *International Herald Tribune*, 9 June 2006, http://www.nytimes.com/iht/2006/06/10/sports/IHT-10cup.html. Another example of the +Teamgeist's impact was the speed of the first of Tomas Rosicky's two goals for the Czech Republic in its game against the USA. Jere Longman, 'U.S. Is Routed by Czech Republic in World Cup', *The New York Times*, 13 June 2006, http://www.nytimes.com/2006/06/13/sports/soccer/13soccer.html
21. Interview with David James, in Rawsthorn, 'A Quest for Perfection'.
22. Interview with Steve Haake, June 2010, in Alice Rawsthorn, 'Design and the World Cup: Best and Worst', *International Herald Tribune*, 27 June 2010, http://www.nytimes.com/2010/06/28/arts/28iht-design28.html; Steve Haake and Simon Choppin, 'Feeling the Pressure: The World Cup's Altitude Factor', *New Scientist*, 4 June 2010, http://www.newscientist.com/article/mg20627635.800-feeling-the-pressure-the-world-cups-altitude-factor.html
23. Ruedi Rüegg died in 2011.
24. 'Ruedi Rüegg', Members, Alliance Graphique Internationale, http://www.a-g-i.org/2147/members/regg.html
25. 'Contesting the Vote: Excerpts From Vice President's Legal Challenge to the Results in Florida', *The New York Times*, 28 November 2000, http://www.nytimes.com/2000/11/28/us/contesting-vote-excerpts-vice-president-s-legal-challenge-results-florida.html
26. Marcia Lausen, *Design for Democracy: Ballot and Election Design* (Chicago, Ill.: University of Chicago Press, 2007), p. 11.
27. Ford Fessenden, 'The 2000 Elections: The Ballot Design', *The New York Times*, 10 November 2000, http://www.nytimes.com/2000/11/10/us/2000-election-ballot-design-candidates-should-be-same-page-experts-say.html; Don Van Natta Jnr and Dana Canedy, 'The 2000 Elections: The Palm Beach Ballot: Democrats Say Ballot's Design Hurt Gore', *The New York Times*, 8 November 2000, http://www.nytimes.com/2000/11/09/us/2000-elections-palm-beach-ballot-florida-democrats-say-ballot-s-design-hurt-gore.html
28. David E. Rosenbaum, 'The 2000 Elections: Florida; State Officials Don't Expect Recount to Change Outcome', *The New York Times*, 8 November 2000, http://www.nytimes.com/2000/11/09/us/2000-elections-florida-state-officials-don-t-expect-recount-change-outcome.html
29. Palm Beach County ended up rejecting some 4.1 per cent of its ballot papers in the 2000 presidential election because of multiple voting, four times more than the national average.
30. 'Pat Buchanan', http://en.wikipedia.org/wiki/Pat_Buchanan; Van Natta Jnr and Canedy, 'The 2000 Elections'.

5 So why is so much design so bad?

1. Jeremy Lewis, *The Life and Times of Allen Lane* (London: Penguin Books, 2006), p. 89.
2. Penny Sparke, *A Century of Car Design* (London: Mitchell Beazley, 2002), pp. 20–21 and 187. The Citroën DS 19 was developed in the early 1950s by a team led by the Italian designer Flaminio Bertoni and the French design engineer André Lefèbvre.

It was unveiled at the Paris Motor Show in 1955. By the end of the first day, Citroën had taken orders for some twelve thousand cars.

3. The French pronunciation of the letters *D* and *S* sounds exactly like *déesse*, the French word for goddess. Roland Barthes, 'The New Citroën', in his *Mythologies* (Frogmore, St Albans: Paladin, 1973), pp. 88–90.
4. Sparke, *A Century of Car Design*, pp. 176–7.
5. Esther da Costa Meyer, 'Simulated Domesticities: Perriand before Le Corbusier', in Mary McLeod, *Charlotte Perriand: An Art of Living* (New York: Harry N. Abrams, 2003), p. 31. In 2006, Charles Harrison was given a Lifetime Achievement Award by the Smithsonian's Cooper-Hewitt, National Design Museum in New York. Megan Gambino, 'Interview with Charles Harrison', 17 December 2008, http://www.smithsonianmag.com/arts-culture/Interview-With-Intelligent-Designer-Charles-Harrison.html
6. Graham Pullin, *Design Meets Disability* (Cambridge, Mass.: The MIT Press. 2009), p. 45.
7. There are cheering exceptions such as spectacles, which have been transformed from corrective devices for people with defective vision into stylish accessories that even those with perfect sight want to wear. The interaction designer Graham Pullin dedicated his 2009 book *'Design Meets Disability* to 'Mr. Cutler and Mr. Gross' of the London opticians Cutler & Gross.
8. Interview with Tom Shakespeare, June 2009, in Alice Rawsthorn, 'Crafting for the Body and Soul', *International Herald Tribune*, 7 July 2009, http://www.nytimes.com/2009/07/06/fashion/06iht-design6.html
9. Interview with Dieter Rams, November 2006, in Alice Rawsthorn, 'Reviving Dieter Rams' Pragmatism', *International Herald Tribune*, 12 November 2006, http://www.nytimes.com/2006/11/12/style/12iht-design13.html
10. Tim Brown, *Change by Design: How Design Thinking Transforms Organizations and Inspires Innovation* (New York: HarperCollins, 2009), p. 25.
11. 'In Praise of . . . Frank Pick', *Guardian*, 17 October 2008.
12. Ken Garland, *Mr Beck's Underground Map* (Harrow: Capital Transport Publishing, 2008).
13. Paul Shaw, *Helvetica and the New York City Subway System: The True (Maybe) Story* (Cambridge, Mass.: The MIT Press, 2010).
14. Suzy Menkes, 'Alexander McQueen, Dark Star of International Fashion', *International Herald Tribune*, 11 February 2010, http://www.nytimes.com/2010/02/12/fashion/12iht-mcqueen.html; Suzy Menkes, 'Galliano's Departure from Dior Ends a Wild Fashion Ride', *International Herald Tribune*, 1 March 2011, http://www.nytimes.com/2011/03/02/business/global/02galliano.html
15. The outcome of such unexpected changes can be positive, as well as negative. To use an example from architecture, the Russian government's sudden decision to increase titanium exports in the mid 1990s prompted such a sharp fall in the price of titanium that the US architect Frank Gehry could suddenly afford to clad the Guggenheim Museum Bilbao with that metal. Titanium is a beautiful material that looks especially lovely in the Basque light, and Gehry's building benefits from it immensely. But often the impact of fluctuating commodity prices is negative. During the late 1960s, the Finnish architect Matti Suuronen sold nearly a hundred of his Futuro mobile homes as prefabricated kits of parts, which were delivered by helicopter and built in the form of an ellipsoid, an ancient shape resembling a flying saucer. So successful was the Futuro that Suuronen designed a larger range of mobile homes to be made in light plastic. But when the cost of that material soared after the 1973 oil price crisis, his new design became too expensive to produce and the Futuro project was scrapped.
16. Steven Heller, *Paul Rand* (London: Phaidon, 1999), pp. 188–9; Michael Bierut, 'The Sins of St Paul', *Observatory: Design Observer*, 31 January 2004, http://observatory.designobserver.com/entry.html?entry=1847
17. 'Brand Identity', Citroën, http://www.citroen.co.uk/home/#/about-us/history
18. The new brand name, which was inspired by one adopted by the house's founding designer Yves Saint Laurent in the 1960s, was introduced solely for the company's ready-to-wear fashion collections. Cassandre's initialled YSL symbol was retained for

other products including perfumes and cosmetics. Yves Saint Laurent coined the brand name 'Saint Laurent Paris' in 1966 for the company's then-new ready-to-wear collection to distinguish it from the traditional haute couture line. The company later adopted a different brand name – Rive Gauche – for ready-to-wear. Hedi Slimane faced fierce criticism for his decision to remove Cassandre's symbol from the ready-to-wear collections. Jess Cartner-Morley, 'Yves Saint Laurent to be Renamed by Creative Director Hedi Slimane', *Guardian*, 21 June 2012, http://www.guardian.co.uk/fashion/2012/jun/21/yves-saint-laurent-renamed

19. After registering his patent for the automatic pop-up toaster in 1919, Charles Strite, a mechanic working in a production plant in Stillwater, Minnesota, co-founded the Waters Genter Company to manufacture the toaster as the Model 1-A-1.
20. George Nelson, *Chairs* (New York: Whitney, 1953), p. 9.
21. Interview with Steve Haake, June 2010, in Alice Rawsthorn 'Design and the World Cup: Best and Worst', *International Herald Tribune*, 27 June 2010, http://www.nytimes.com/2010/06/28/arts/28iht-design28.html?ref=arts; Steve Haake and Simon Choppin, 'Feeling the Pressure: The World Cup's Altitude Factor', *New Scientist*, 4 June 2010, http://www.newscientist.com/article/mg20627635.800-feeling-the-pressure-the-world-cups-altitude-factor.html

6 **Why everyone wants to 'do an Apple'**

1. 'The First iMac Introduction', uploaded on to YouTube by peestandingup on 30 January 2006, http://www.youtube.com/watch?v=0BHPtoTctDY
2. Leander Kahney, *Inside Steve's Brain* (New York: Portfolio, 2008), pp. 15 and 16.
3. David Streitfield, 'Jobs Steps Down at Apple, Saying He Can't Meet Duties', *The New York Times*, 24 August 2011, http://www.nytimes.com/2011/08/25/technology/jobs-stepping-down-as-chief-of-apple.html
4. 'Steve Jobs Shrines around the World – in Pictures', *Guardian*, 6 October 2011, http://www.guardian.co.uk/technology/gallery/2011/oct/06/steve-jobs-apple-shrines-world#/?picture=379996403&index=0
5. Nick Bilton, 'The Rise of the Fake Apple Store', *The New York Times: Bits Blog*, 20 July 2011, http://bits.blogs.nytimes.com/2011/07/20/the-rise-of-the-fake-apple-store/
6. 'A Genius Departs', *Economist*, 8 October 2011, http://www.economist.com/node/21531530
7. Alex Williams, 'Short Sainthood for Steve Jobs', *The New York Times*, 2 November 2011, http://www.nytimes.com/2011/11/03/fashion/the-steve-jobs-backlash.html. There were, inevitably, some dissenters, including 'MikeinOhio', who posted 'Was Steve Jobs a Good Man or an Evil Corporate CEO and Wall Street Shill?' on the Occupy Wall Street website on 6 October 2011, but such critics were in the minority: http://occupywallst.org/forum/was-steve-jobs-a-good-man-or-an-evil-corporate-ceo/
8. Alison Weir, *Henry VIII: King and Court* (London: Jonathan Cape, 2001), pp. 186–94.
9. Alison Kelly (ed.), *The Story of Wedgwood* (1962; London: Faber and Faber, 1975), p. 18.
10. Elizabeth Templetown and her husband, Clotworthy Upton, who was given the title Baron Templetown, commissioned Robert Adam to remodel their country home in 1783. It was a sixteenth-century castle built on the site of a medieval fort at Templepatrick in County Antrim, Northern Ireland, that the Upton family had bought in the early seventeenth century and renamed Castle Upton.
11. Novalis was the nom de plume of Georg Philipp Friedrich Freiherr von Hardenberg. *Novalis: Philosophical Writings* (1798; Albany: State University of New York Press, 1977), p. 111.
12. Herbert Read and Bernard Rackham, *English Pottery* (London: Ernest Benn, 1924), p. 26.
13. Andrea Gleiniger, *The Chair No. 14 by Michael Thonet* (Frankfurt am Main: form, 1998), pp. 7–17.
14. Alexander von Vegesack, *Thonet: Classic Furniture in Bent Wood and Tubular Steel* (London: Hazar, 1996), pp. 32–4 and 116.
15. Interview with Konstantin Grcic, September 2008.
16. Von Vegesack, *Thonet*, pp. 34 and 36–7.

17. *Ibid.*, p. 109.
18. Siegfried Gronert, 'From Material to Model: Wagenfeld and the Metal Workshops at the Bauhaus and the Bauhochschule in Weimar', in Beate Manske (ed.), *Wilhelm Wagenfeld (1900–1990)* (Ostfildern-Ruit: Hatje Cantz, 2000), pp. 12–18.
19. Dieter Rams, *Less but Better* (Hamburg: Jo Klatt Design+Design, 1995), pp. 9–10.
20. *Ibid.*, pp. 15–16.
21. Bernd Polster, 'Kronberg Meets Cupertino: What Braun and Apple Really Have in Common', in Sabine Schulze and Ina Grätz (eds), *Apple Design* (Ostfildern: Hatje Cantz, 2011), pp. 68–9.
22. Keiko Ueki-Polet and Klaus Kemp (eds), *Less and More: The Design Ethos of Dieter Rams* (Berlin: Gestalten, 2009), p. 115.
23. Rams, *Less but Better*, p. 19.
24. Klaus Kemp, 'Dieter Rams, Braun, Vitsoe and the Shrinking World', in Ueki-Polet and Kemp, *Less and More*, p. 467.
25. Sophie Lovell, *Dieter Rams: As Little Design as Possible* (London: Phaidon, 2011), p. 235.
26. Rams, *Less but Better*, pp. 57–8.
27. Lovell, *Dieter Rams*, p. 239.
28. Klaus Kemp, 'Dieter Rams, Braun, Vitsoe and the Shrinking World', in Ueki-Polet and Kemp, *Less and More*, p. 465.
29. *Ibid.*, p. 467.
30. Lovell, *Dieter Rams*, p. 13.
31. Steve Jobs introducing the first iMac on 6 May 1998 in the Flint Auditorium at De Anza Community College in Cupertino, California. 'The First iMac Introduction' uploaded on to YouTube by peestandingup on 30 January 2006, http://www.youtube.com/watch?v=0BHPtoTctDY
32. Walter Isaacson, *Steve Jobs* (London: Little, Brown, 2011), p. 126.
33. Steve 'Woz' Wozniak was five years older than Steve Jobs and the star student in the pioneering electronics class at Homestead High School in Silicon Valley. They had both attended the school, but at different times. Woz's younger brother had been on the school swim team with Jobs. *Ibid.*, pp. 21–5.
34. Mark Frauenfelder, *The Computer* (London: Carlton Books, 2005), p. 135.
35. Isaacson, *Steve Jobs*, pp. 73 and 83.
36. David Sheff, 'Playboy Interview: Steven Jobs', *Playboy*, 1 February 1985, http://www.txtpost.com/playboy-interview-steven-jobs/
37. Ina Grätz, 'Stylectrical: On Electro-Design That Makes History', in Schulze and Grätz, *Apple Design*, p. 14.
38. Isaacson, *Steve Jobs*, p. 126.
39. Thomas Wagner, 'Think Different! Users and Their Darlings: On Apples, Machines, Interfaces, Magic, and the Power of Design', in Schulze and Grätz, *Apple Design*, p. 35.
40. Isaacson, *Steve Jobs*, pp. 133 and 186.
41. *Ibid.*, pp. 220–21.
42. James B. Stewart, 'How Jobs Put Passion into Products', *The New York Times*, 7 October 2011, http://www.nytimes.com/2011/10/08/business/how-steve-jobs-infused-passion-into-a-commodity.html
43. Isaacson, *Steve Jobs*, p. 350.
44. Lovell, *Dieter Rams*, p. 13.
45. 'New iPad 2: Thinner, Lighter, Faster', uploaded on to YouTube by marvinsc on 2 March 2011, http://www.youtube.com/watch?v=iy017Af_V0o
46. Isaacson, *Steve Jobs*, pp. 458–9.
47. *Ibid.*, pp. 344–5.
48. Malcolm Gladwell, 'The Tweaker: The Real Genius of Steve Jobs', *New Yorker*, 14 November 2011, http://www.newyorker.com/reporting/2011/11/14/111114fa_fact_gladwell
49. Steve Jobs, 'Apple's One-Dollar-A-Year Man', *Fortune*, 24 January 2000, http://money.cnn.com/magazines/fortune/fortune_archive/2000/01/24/272277/
50. 'Apple Steve Jobs on Design', a 2002 corporate film clip in which Steve Jobs and

Jonathan Ive discuss Apple's approach to design, uploaded on to YouTube by DirkBeveridge1340 on 16 October 2010, http://www.youtube.com/watch?v=sPfJQmpg5zk

51. Jobs, 'Apple's One-Dollar-A-Year Man'.
52. Isaacson, *Steve Jobs*, p. 389.
53. *Ibid.*, p. 373.
54. Jobs, 'Apple's One-Dollar-A-Year Man'.
55. Brad King and Farhad Manjoo, 'Apple's "Breakthrough" iPod', *Wired*, 23 October 2001, http://www.wired.com/gadgets/miscellaneous/news/2001/10/47805
56. Jacqui Cheng, 'Steve Jobs on MobileMe: The Full E-mail', *Ars Technica* (August 2008), http://arstechnica.com/apple/news/2008/08/steve-jobs-on-mobileme-the-full-e-mail.ars
57. Isaacson, *Steve Jobs*, pp. 344–5.
58. A report by Charles Duhigg and David Barboza published in *The New York Times* in January 2012 into the working conditions for the employees of some of Apple's Chinese subcontractors sparked a controversy about Apple's ethical record. Charles Duhigg, David Barboza, 'In China, Human Costs Are Built into an iPad', *The New York Times*, 25 January 2012, http://www.nytimes.com/2012/01/26/business/ieconomy-apples-ipad-and-the-human-costs-for-workers-in-china.html; John Cassidy, 'Rational Irrationality: How Long Will the Cult of Apple Endure?', *New Yorker Blog*, 20 March 2012, http://www.newyorker.com/online/blogs/johncassidy/2012/03/how-long-will-the-cult-of-apple-last-for.html
59. *Objectified*, directed by Gary Hustwit, produced by Plexi Productions, Swiss Dots (2009).
60. 'Apple Steve Jobs on Design', uploaded on to YouTube. The American designer Charles Eames said, 'The details are not the details. They make the product' when narrating a film on the ECS contract storage system he designed with his wife, Ray Eames: 'Charles Eames', Art Directors Club, http://www.adcglobal.org/archive/hof/1984/?id=245

7 Why design is not – and should never be confused with – art

1. Bruno Munari, *Design as Art* (1966; London: Penguin Books, 2008), p. 25.
2. Richard Morphet (ed.), *Richard Hamilton* (London: Tate Gallery Publications, 1992), p. 164.
3. As Sophie Lovell pointed out in her book on Dieter Rams, he did not design the HT 2 toaster, which was the work of a Braun colleague, Reinhold Weiss. However, Rams was in overall charge of Braun's design team during its development. *Dieter Rams: As Little Design as Possible* (London: Phaidon, 2011), p. 293.
4. Morphet, *Richard Hamilton*, p. 164. Richard Hamilton remade *Toaster* using chromium-plated steel, rather than aluminum, as in the original piece.
5. Plato, *The Republic* (1955; London: Penguin Classics, 2007), p. 340.
6. Giorgio Vasari, *Lives of the Artists: Volume I* (1550; London: Penguin Books, 1987), pp. 256–7.
7. *Ibid.*, pp. 232–40.
8. Herbert Read, *Art and Industry* (London: Faber and Faber, 1934), p. 9.
9. The Académie des Beaux-Arts was founded in Paris in 1648 as the Académie Royale de Peinture et de Sculpture when Louis XIV was a child at the behest of his chief minister, Cardinal Mazarin. It was restructured under the aegis of Jean-Baptiste Colbert in 1663, and used as part of his strategy of 'glorifying' Louis's reputation under Charles Le Brun as director. Peter Burke, *The Fabrication of Louis XIV* (New Haven, Conn.: Yale University Press, 1992), pp. 50–51.
10. Celina Fox, *The Arts of Industry in the Age of Enlightenment* (New Haven, Conn.: Yale University Press, 2009), p. 453.
11. Fiona MacCarthy, *The Last Pre-Raphaelite: Edward Burne-Jones and the Victorian Imagination* (London: Faber and Faber, 2011), p. 33.
12. Magdalena Bushart, 'It Began with a Misunderstanding: Feininger's Cathedral and the Bauhaus Manifesto', in Bauhaus-Archiv Berlin, Stiftung Bauhaus Dessau and Klassik Stiftung Weimar, *Bauhaus: A Conceptual Model* (Ostfildern: Hatje Cantz, 2009), pp. 30–32.

13. Nicholas Fox Weber and Pandora Tabatabai Asbaghi, *Anni Albers* (New York: Guggenheim Museum Publications, 1999), p. 155.
14. Ulrich Herrmann, 'Practice, Program, Rationale: Johannes Itten and the Preliminary Course at the Weimar Bauhaus', in Bauhaus-Archiv Berlin et al., *Bauhaus*, pp. 68–70.
15. Sibyl Moholy-Nagy, *Moholy-Nagy: Experiment in Totality* (New York: Harper & Brothers, 1950), pp. 32–47.
16. Franz Schulze, *Philip Johnson: Life and Work* (New York: Alfred A. Knopf, 1994), p. 54.
17. Philip Johnson, *Machine Art* (New York: Museum of Modern Art, 1934).
18. Kai-Uwe Hemken, 'Cultural Signatures: László Moholy-Nagy and the "Room of Today"', in Ingrid Pfeiffer and Max Hollein (eds), *László Moholy-Nagy Retrospective* (Munich: Prestel, 2009), pp. 168–71; Hans Ulrich Obrist (ed.), *A Brief History of Curating* (Zurich: J. R. P. Ringier, 2008).
19. Alexander Dorner, catalogue for Herbert Bayer exhibition at the London Gallery, 8 April to 1 May 1937, p. 6.
20. *Time*, 31 October 1949.
21. Steven Heller, *Paul Rand* (London: Phaidon, 1999), p. 149.
22. John Neuhart, Marilyn Neuhart and Ray Eames, *Eames Design: The Work of the Office of Charles and Ray Eames* (London: Thames & Hudson, 1989), pp. 285–91.
23. *Ibid.*, pp. 440–41; 'Powers of Ten (1977)', uploaded on to YouTube by Eames Office on 26 August 2010, http://www.youtube.com/watch?v=0fKBhvDjuy0
24. László Moholy-Nagy, *Vision in Motion* (Chicago, Ill.: Paul Theobald, 1947), p. 42.
25. György Kepes, *Language of Vision* (1944; New York: Dover Publications, 1995).
26. Emily King, *Robert Brownjohn: Sex and Typography* (New York: Princeton Architectural Press, 2005), pp. 194–223.
27. Achille and Pier Giacomo Castiglioni worked together from the opening of their Milan studio in 1944 until the latter's death in 1968. Their brother Livio worked with them until he left to open his own office in 1952.
28. Fulvio and Napoleone Ferrari, *The Furniture of Carlo Mollino* (London: Phaidon, 2006); Chris Dercon (ed.), *Carlo Mollino: Maniera Moderna* (Cologne: Walther König, 2011).
29. Yves Saint Laurent's autumn/winter 1966 haute couture women's wear collection was inspired by Andy Warhol's art and dedicated to him. Warhol had met and befriended Saint Laurent and his then-lover and business partner Pierre Bergé during a visit to Paris in the summer of 1966, and the three men remained close friends until the artist's death in 1987.
30. Katya García-Antón, 'Performative Shifts in Art and Design', in Katya García-Antón, Emily King and Christian Brandle, *Wouldn't It Be Nice . . . Wishful Thinking in Art and Design* (Geneva: Centre d'Art Contemporain de Genève, 2007), p. 57.
31. Peter Weiss (ed.), *Alessandro Mendini: Design and Architecture* (Milan: Electa, 2001), p. 79.
32. Glenn Adamson and Jane Pavitt (eds), *Postmodernism: Style and Subversion, 1970–1990* (London: V&A Publishing, 2011), p. 15.
33. Another of Alessandro Mendini's Redesign pieces was his take on the Dutch designer Thomas Gerrit Rietveld's 1932–4 Zig Zag chair with its wooden spine, which extended to create the shape of a cross. Weiss, *Alessandro Mendini*, pp. 80–81.
34. Barbara Radice, *Ettore Sottsass: A Critical Biography* (New York: Rizzoli International, 1993), pp. 212–16.
35. Robert Venturi, Denise Scott Brown and Steven Izenour, *Learning from Las Vegas: The Forgotten Symbolism of Architectural Form* (1972; Cambridge, Mass.: The MIT Press, 1977).
36. Emily King (ed.), *Designed by Peter Saville* (London: Frieze, 2003).
37. Wolfgang Tillmans in Heike Munder (ed.), *Peter Saville Estate 1–127* (Zurich: Migros Museum für Gegenwartskunst Zürich and JRP|Ringier, 2007).
38. Raymond Williams, 'Culture', in his *Keywords: A Vocabulary of Culture and Society* (1976; London: Fontana, 1983), p. 27.
39. Nicolas Bourriaud first mentioned the term 'Relational Aesthetics' in the catalogue for 'Traffic', an exhibition he curated in 1996 at the CAPC Musée d'Art Contemporain de Bordeaux. He then expanded upon the concept in a book, *Esthétique Relationnelle*'

(or 'Relational Aesthetics') (Dijon: Les Presses du Réel, 1998).
40. Ai Weiwei was co-director of the Gwangju Design Biennale 2011, entitled 'Design Is Design Is Not Design'. His section of the biennale was the 'Un-Named' exhibition featuring design projects which would not conventionally be considered to have been 'designed', such as a computer virus and an activists' plan for a protest in Cairo during the Arab Spring uprising in 2011; http://gb.or.kr/?mid=main_eng
41. 'Common Cause: Elizabeth Schambelan Talks with Carolyn Christov-Bakargiev about Documenta 13', *Artforum* (May 2012).
42. Phillips later changed its name to Phillips de Pury, then back to Phillips.
43. Judd Tully, 'Marc Newsons Falter at Lackluster Phillips Design Auction', *Blouin Art Info*, 10 June 2010, http://www.artinfo.com/news/story/34882/marc-newsons-falter-at-lackluster-phillips-design-auction/
44. Calvin Tomkins, *Duchamp: A Biography* (New York: Henry Holt, 1996); Thierry de Duve, 'Echoes of the Readymade: Critique of Pure Modernism', in Martha Buskirk and Mignon Nixon (eds), *The Duchamp Effect: Essays, Interviews, Round Table* (Cambridge, Mass.: The MIT Press, 1996), pp. 93–129.
45. The concept of the film-maker as an auteur, which literally translates as 'author', was a central tenet of the Nouvelle Vague, or 'new wave' of French film-makers in the 1950s. The auteur was defined by the critic and film-maker François Truffaut in an essay for the January 1954 issue of the journal *Cahiers du Cinéma*, entitled *'Une Certaine Tendance du Cinema Français'* ('A Certain Tendency in French Cinema') as a film-maker who treated his or her work as a medium of self-expression by imbuing it with his or her sensibility, ideas and principles. Truffaut referred to the rise of the auteur in French cinema as '*La politique* (or policy) *des auteurs*', but the American film critic Andrew Sarris gave it the new name of 'the auteur theory'. David A. Cook, *A History of Narrative Film* (New York: W.W. Norton, 1990), p. 552.
46. Bill Moggridge and Olga Viso's 'Directors' Foreword', in Andrew Blauvelt and Ellen Lupton (eds), *Graphic Design: Now in Production* (Minneapolis, Minn.: Walker Art Center, 2011), p. 6.
47. Rick Poynor, *No More Rules: Graphic Design and Postmodernism* (London: Laurence King, 2003).
48. Blauvelt and Lupton, *Graphic Design.*
49. Irma Boom (ed.), *Irma Boom: Biography in Books* (Amsterdam: University of Amsterdam Press, 2010), pp. 512–14.
50. Metahaven and Marina Vishmidt, *Uncorporate Identity* (Baden: Lars Müller Publishers with the Jan van Eyck Academie, Maastricht, 2010).
51. The term 'critical design' was first used in Anthony Dunne's *Hertzian Tales: Electronic Products, Aesthetic Experience and Critical Design* (London: Royal College of Art Computer Related Design Research Studio, 1999). Dunne and Fiona Raby refined the concept through their design practice, teaching, writing and curation, notably in the 2007 exhibition 'Designing Critical Design' at Z33 in Hasselt, Belgium, where their work was exhibited alongside that of the Dutch designer Jurgen Bey and the Spanish designer Martí Guixé: 'Nr. 15 Designing Critical Design', Z33, http://www.z33.be/en/projects/nr-15-designing-critical-design
52. Christian Brandle, 'Dunne & Raby and Michael Anastassiades', in García-Antón, King and Brandle, *Wouldn't It Be Nice*, pp. 153–66.
53. Martí Guixé, http://www.guixe.com; Martí Guixé (ed.), *Libre de Contexte, Context Free* (Basel: Birkhäuser, 2003).
54. 'Julia: Cow benches, 2005', Julia Lohmann, http://www.julialohmann.co.uk/work/gallery/cow-benches/
55. Coralie Gauthier (ed.), *Mathieu Lehanneur* (Berlin: Gestalten, 2012).
56. Donald Judd, 'It's Hard to Find a Good Lamp' (1993), http://www.juddfoundation.org/furniture/essay.htm This essay was originally published in the exhibition catalogue *Donald Judd Furniture* (Rotterdam: Museum Boijmans van Beuningen,1993).
57. Brandle, 'Dunne & Raby and Michael Anastassiades', in García-Antón, King and Brandle, *Wouldn't It Be Nice*, p. 154.
58. Charles Eames was interviewed by Yolande Amic in 1972 to mark the exhibition

'Qu'est-ce Que le Design?' at the Musée des Arts Décoratifs in Paris. 'Design Q&A with Charles Eames (1972)', uploaded on to YouTube by myuichirou on 13 February 2010, http://www.youtube.com/watch?v=z8qs5-BDXNU&feature=BFa&list=PL0054AB7E04F53D0E&lf=results_video

8 Sign of the times

1. Raymond Loewy, *Industrial Design* (London: Faber and Faber, 1979), p. 32.
2. Walter Isaacson, *Steve Jobs* (London: Little, Brown, 2011), p. 63.
3. *Ibid.*, p. 80.
4. Andrew Hodges, *Alan Turing: The Enigma* (1983; London: Vintage, 2012), pp. 487–96.
5. Apple, http://www.apple.com
6. Thomas K. Grose, 'Naughty But Nice', *Time*, 24 October 1999, http://www.time.com/time/magazine/article/0,9171,33155,00.html
7. Ann Summers, http://www.annsummers.com
8. Robert Graves, *Greek Myths* (1955; London: Cassell, 1991), pp. 22–4. The olive branch also appeared as a symbol of peace in Aristophanes' fifth-century BC play *Peace*, and later in Virgil's epic poem *The Aeneid*.
9. Robert Recorde, *The Whetstone of Witte* (1557; Mountain View, Calif.: Creative Commons, 2009).
10. Captain Charles Johnson, *A General History of the Robberies and Murders of the Most Notorious Pyrates* (1724; London: Conway Maritime Press, 2002).
11. *Typographica*, 3 and 4 (1961). *Typographica* was a design magazine specializing in typography, founded by the British designer, editor and writer Herbert Spencer in 1949, when he was only twenty-five years old, and published until 1967.
12. Rick Poynor (ed.), *Communicate: Independent British Graphic Design since the Sixties* (London: Barbican Art Gallery and Laurence King, 2004), p. 81.
13. Interview with Margaret Calvert in 2005.
14. Benno Wissig's 1967 signage scheme for Amsterdam's Schiphol Airport used highly legible lettering and strict colour-coding to produce a clear and coherent information system. To minimize confusion, Wissig banned any other signs in his chosen colours of yellow and green from the airport. Even Hertz, the hire-car company, had to ditch its customary yellow signs. Wissig's original design scheme for Schiphol was revised in 1991 by Paul Mijksenaar, who has since maintained it with skill and sensitivity.
15. Paola Antonelli, '@ at MoMA', Inside/Out: A MoMA/MoMA PS1 Blog, 22 March 2010, http://www.moma.org/explore/inside_out/2010/03/22/at-moma
16. Chris Messina, 'how do you feel about using . . .', Twitter, https://twitter.com/#!/chrismessina/status/223115412
17. Twitter Fan Wiki, http://twitter.pbworks.com/w/page/1779812/Hashtags
18. Randal C. Archibald, 'California Fires Force 500,000 from Homes', *The New York Times*, 23 October 2007, http://www.nytimes.com/2007/10/23/us/23cnd-fire.html
19. Nate Ritter, http://nateritter.com/
20. Antonelli, '@ at MoMA'.
21. The Nazis' homosexual prisoners remained incarcerated in concentration camps for considerably longer than many of their fellow inmates. Homosexuality was a criminal offence in pre-war Germany under the notorious Paragraph 175, which remained in force in West Germany after the Second World War until its repeal in 1969. Only then, some twenty-four years after the end of the war, were the Nazi regime's surviving homosexual prisoners released.
22. 'The CND Symbol', CND, http://www.cnduk.org/about/item/435
23. The average number of corporate symbols seen by a typical Western consumer each day is generally estimated at 3,000, although it is not clear how this number has been calculated or how accurate it is.
24. BMW Group, http://www.bmwgroup.com; Fiat, http://www.fiat.com; Officina Profumo Farmaceutica di Santa Maria Novella, http://www.smnovella.it
25. 'Our Heritage', Kellogg's, http://www.kelloggcompany.com/company.aspx?id=39
26. Stephen Banham, *Characters: Cultural Stories Revealed through Typography* (Melbourne: Thames & Hudson, 2011), p. 70.

27. '125 Years of Coca-Cola logos', Coca-Cola, http://www.coca-cola.co.uk/125/history-of-coca-cola-logo.html
28. Hermès, http://www.hermes.com
29. Manchester United Official Website, http://www.manutd.com
30. 'Citroën's History', Citroën, http://www.citroen.co.uk/home/#/about-us/history/brand-identity/
31. 'Symbols of NASA', http://www.nasa.gov/audience/forstudents/5-8/features/symbols-of-nasa.html
32. Steven Heller, *Paul Rand* (London: Phaidon Press, 1999), pp. 149–73.
33. Steve Jobs recalled working with Paul Rand on the development of the NeXT identity, and Rand's love for eating at Gold's Delicatessen in Newport, Connecticut, in a video interview filmed in 1993. '1993 Steve Jobs interview about working with Paul Rand', http://www.paul-rand.com/foundation/video_stevejobs_interview/#.UECi2xz5Mbo
34. Heller, *Paul Rand*, p. 156.
35. Alan Hess, 'The Origins of McDonald's Golden Arches', *Journal of the Society of Architectural Historians*, 45:1 (March 1986), pp. 60–67.
36. 'About FedEx', FedEx, http://about.van.fedex.com/taxonomy/term/2938
37. Nike's 'swoosh' was designed by Carolyn Davidson when she was studying graphic design at Portland State University near Nike's headquarters. Nike's co-founder Phil Knight taught an accounting class there to make ends meet while he and his partner, Bill Bowerman, were building up the business, which was then called Blue Ribbon Sports Inc. They had one full-time employee, Jeff Johnson, whom Knight had met while studying for his MBA at Stanford University, and it was he who invented the name Nike in 1971. When Knight noticed Davidson working on an assignment during one of his visits to Portland State, he asked her to design a logo for the renamed company. The first time Knight saw the swoosh he said, 'I don't love it, but it will grow on me.' Nike paid Davidson $35 for designing the symbol, but also gave her a job. 'History & Heritage', Nike, http://nikeinc.com/pages/history-heritage
38. Isaacson, *Steve Jobs*, pp. 220–21.
39. 'The Michelin Man', Michelin Corporate, http://www.michelin.com/corporate/EN/group/michelin-man
40. 'Nunc est Bibendum' is the title of a poem written by Horace in 23 BC, which translates from Latin as 'Now is the time for drinking'.
41. Frank Olinsky, http://www.frankolinsky.com/mtvstory1.html
42. 'Cook Books, Food Writing & Recipes', Penguin Books, http://www.penguin.co.uk/static/cs/uk/0/penguin_food/index.html
43. Interview with Bruce Mau, February 2007, in Alice Rawsthorn, 'The New Corporate Logo: Dynamic and Changeable Are All the Rage', *International Herald Tribune*, 11 February 2007, http://www.nytimes.com/2007/02/11/style/11iht-design12.html
44. British Airways replaced half of the multi-ethnic tail fins with a more conventional Union Jack-inspired design scheme in 1999, and announced in 2001 that it was to replace the rest. 'R.I.P. British Airways' Funky Tailfins', BBC News, 11 May 2001, http://news.bbc.co.uk/1/hi/uk/1325127.stm
45. 'Avant-garde Shoreditch E1', http://www.telfordhomes.plc.uk/avantgardetower/
46. Occupy Wall Street, http://www.occupywallst.org/

9 When a picture says more than words

1. Janet Abrams, 'Muriel Cooper', AIGA website to mark the award of the 1994 AIGA Medal to Cooper, http://www.aiga.org/medalist-murielcooper/; reprinted from 'Flashback: Muriel Cooper's Visible Wisdom', *I.D. Magazine* (September–October 1994).
2. 'Charles Booth (1840–1914) – a Biography', Charles Booth Online Archive, http:// booth.lse.ac.uk/static/a/2.html
3. Charles Booth continued to campaign against poverty throughout his life, as did many of his researchers. Clara Collet became an authority on the rights of women workers. Beatrice Webb made an important contribution to the foundation of the welfare state. Booth himself lobbied for the introduction of state pensions to alleviate the hardship of the elderly.

4. Deborah McDonald, *Clara Collet, 1860–1948: An Educated Working Woman* (London: Routledge, 2004).
5. Charles Booth, *Life and Labour of the People in London: Volume 1* (London: Macmillan, 1902), pp. 33–62.
6. 'Poverty Maps of London', Charles Booth Online Archive, http://booth.lse.ac.uk/static/a/4.html#i
7. 'Inquiry into the Life and Labour of the People in London (1886–1903)', Charles Booth Online Archive, http://booth.lse.ac.uk/static/a/3.html#ii
8. Edward R. Tufte, *The Visual Display of Quantitative Information* (Cheshire, Conn.: Graphics Press, 2001), p. 20.
9. Edward Tufte, *Beautiful Evidence* (Cheshire, Conn.: Graphics Press, 2006), pp. 20–21 and 97–103.
10. Tufte, *The Visual Display of Quantitative Information*, pp. 32–4.
11. Tufte, *Beautiful Evidence*, pp. 22–3.
12. Tufte, *The Visual Display of Quantitative Information*, p. 24.
13. Tufte, *Beautiful Evidence*, pp. 134–5.
14. Tufte, *The Visual Display of Quantitative Information*, pp. 40–41.
15. Otto Neurath, *From Hieroglyphics to Isotype: A Visual Autobiography* (London: Hyphen Press, 2010), pp. 23–39 and 69–72.
16. Ed Annink and Max Bruinsma (eds), *Gerd Arntz: Graphic Designer* (Rotterdam: 010 Publishers, 2010).
17. The Isotype team left Vienna in the mid 1930s when the Nazis came to power. Otto Neurath and Marie Reidemeister fled to the Netherlands with Gerd Arntz, who remained there and severed his links with Isotype after they left for Britain in 1940. They set up the Isotype Institute in Oxford and assembled a new team of researchers there. After Neurath died in 1945, Reidemeister continued their work at the institute, notably by undertaking public education projects in Africa.
18. Lella Secor Florence, *Our Private Lives: America and Britain* (London: George G. Harrap, 1944), pp. 4–6.
19. Marie Neurath and Robert S. Cohen (eds), *Otto Neurath: Empiricism and Sociology* (Dordecht: D. Reidel, 1973), p. 217.
20. Ken Garland, *Mr Beck's Underground Map* (Harrow: Capital Transport Publishing, 2008), p. 15.
21. *Ibid.*, p. 17.
22. *Ibid.*, p. 19.
23. *Ibid.*, p. 35.
24. *Ibid.*, p. 19.
25. *Ibid.*, p. 44.
26. *Ibid.*, pp. 21 and 26.
27. *Ibid.*, pp. 50–61.
28. *Ibid.*, p. 67.
29. Born in Milan in 1931, Massimo Vignelli studied architecture there and in Venice, and visited the United States on a research fellowship in the late 1950s. He moved to New York in 1965 to run the office of Unimark International, a design group he had co-founded earlier that year with Ralph Eckerstrom, former design director of the Container Corporation of America. Herbert Bayer, the former Bauhaüsler, acted as a consultant to Unimark. In 1966, Unimark was appointed to advise the New York City Transit Authority on design at the suggestion of the Museum of Modern Art, New York. Vignelli and his wife Leila co-founded Vignelli Associates in 1971, where their clients have included IBM, Knoll, Bloomingdales and American Airlines.
30. Paul Shaw, *Helvetica and the New York City Subway System: The True (Maybe) Story* (Cambridge, Mass.: The MIT Press, 2010), p. 125. The New York-based graphic designer Michael Bierut posted a tribute to the purist design of Massimo Vignelli's subway map on the Design Observer website in September 2010: 'Mr. Vignelli's Map', http://observatory.designobserver.com/entry.html?entry=2647
31. Michael Kimmelman 'The Grid at 200: Lines That Shaped Manhattan', *The New York Times*, 2 January 2012, http://www.nytimes.com/2012/01/03/

arts/design/manhattan-street-grid-at-museum-of-city-of-new-york.html
32. 'Massimo Vignelli and His 1972 NY Subway Map', uploaded on to YouTube by swissdots on 21 June 2008, http://www.youtube.com/watch?v=uhMKHXLBZrc
33. 'The Data Deluge: Businesses, Governments and Society are Only Starting to Tap Its Vast Potential', *Economist*, 27 February 2010, http://www.economist.com/node/15579717
34. Gordon E. Moore predicted in a 1965 paper published in *Electronics Magazine* that the processing power and storage capacity of computer chips would double, or their prices would halve, roughly every two years for at least the next decade. His principle is known in the tech industry as 'Moore's law'. Intel subsequently reduced the time frame from two years to eighteen months, and computing power has continued to increase at a similar rate since passing Moore's original 1975 deadline.
35. 'Cisco Visual Networking Index: Forecast and Methodology, 2011–2016', Cisco Systems, http://www.cisco.com/en/US/solutions/collateral/ns341/ns525/ns537/ns705/ns827/white_paper_c11-481360_ns827_Networking_Solutions_White_Paper.html
36. 'Data, Data Everywhere: A Special Report on Managing Information', *Economist*, 27 February 2010, http://www.economist.com/node/15557443
37. Achim Borchardt-Hume (ed.), *Albers and Moholy-Nagy: From the Bauhaus to the New World* (London: Tate Publishing, 2006), pp. 69–70. The Light Space Modulator was originally known as 'Light Prop for an Electric Stage' when Moholy-Nagy first used it in Berlin; *ibid.*, pp. 95–6.
38. Judith Wechsler, 'György Kepes', in György Kepes (ed.), *György Kepes: The MIT Years, 1945–1977* (Cambridge, Mass.: The MIT Press, 1978), pp. 11–19.
39. György Kepes, *Language of Vision* (1944; New York: Dover Publications, 1995); Abrams, 'Muriel Cooper'.
40. Honor Beddard and Douglas Dodds, *Digital Pioneers* (London: V&A Publishing, 2009).
41. Desmond Paul Henry, http://www.desmondhenry.com/index.html
42. Interview with Elaine O'Hanrahan (daughter of Desmond Paul Henry), February 2011, in Alice Rawsthorn, 'When Desmond Paul Henry Traded His Pen for a Machine', *International Herald Tribune*, 27 February 2011, http://www.nytimes.com/2011/02/28/arts/28iht-design28.html
43. 'Visualization: A Way to See the Unseen', National Science Federation, http://www.nsf.gov/about/history/nsf0050/pdf/visualization.pdf
44. Bruce H. McCormick, Thomas A. DeFanti and Maxine D. Brown, 'Visualization in Scientific Computing', *Computer Graphics*, 21:6 (November 1987), http://www.evl.uic.edu/files/pdf/ViSC-1987.pdf
45. An example of the early visualizations of Internet traffic is the Opte Project developed by Darren Lyon from October 2003 onwards; http://www.opte.org/
46. 'Overview: A Short Introduction to the Processing Software and Projects from the Community', Processing, http://www.processing.org/about/
47. Ben Fry, 'Humans vs Chimps', http://benfry.com/humansvschimps/
48. 'The Weekender', http://www.mta.info/weekender.html. Massimo Vignelli had the last laugh after the furore over his 1972 New York Subway Map, when the MTA invited him to reinterpret that design scheme for 'The Weekender'. He agreed to do so on condition that it was described as a 'diagram', not as a map. One of the amendments Vignelli made to the original was to erase the parks, including Central Park.
49. The Blue Brain project began in 2005 with an agreement between the École Polytechnique Fédérale de Lausanne and IBM, which supplied the BlueGene/L supercomputer acquired by EPFL to build the virtual brain. Creating each simulated neuron requires the equivalent processing power of a laptop computer, and there are billions of neurons in the whole brain. A human cortex may have as many as two million columns, each of which consists of some hundred thousand neurons. The speed with which the project is completed will be determined by the pace of developments in supercomputing technology. So far those advances have been faster than originally expected; http://bluebrain.epfl.ch/
50. Joost Grootens, *I swear I use no art at all: 10 years, 100 books, 18,788 pages of book design* (Rotterdam: 010 Publishers, 2010).

51. Interview with Joost Grootens, January 2011, in Alice Rawsthorn, 'Designing Books for a Digital Age', *International Herald Tribune*, 30 January 2011, http://www.nytimes.com/2011/01/31/arts/31iht-design31.html

10 It's not that easy being green

1. György Kepes, 'Comments on Art', in Abraham H. Maslow (ed.), *New Knowledge in Human Values* (New York: Harper & Brothers, 1959), p. 91.
2. The phrase 'Make it a green peace' was coined at that meeting of the Don't Make a Wave Committee by the ecologist Bill Darnell.
3. 'Amchitka: The Founding Voyage', Greenpeace International, http://www.greenpeace.org/international/en/about/history/amchitka-hunter/
4. Victoria Finlay, *Colour* (London: Sceptre, 2002), pp. 289–98.
5. Interview with Michael Braungart, March 2010, in Alice Rawsthorn, 'The Toxic Side of Being, Literally, Green', *International Herald Tribune*, 4 April 2010, http://www.nytimes.com/2010/04/05/arts/05iht-design5.html
6. 'Sesame Street: Kermit Sings Being Green', uploaded on to YouTube by Sesame Street on 11 December 2008, http://www.youtube.com/watch?v=51BQfPeSK8k
7. *The New York Times Magazine* published a haunting series of photographs by Pieter Hugo of the Agbogbloshie digital dump in 2010. 'A Global Graveyard for Dead Computers in Ghana', http://www.nytimes.com/slideshow/2010/08/04/magazine/20100815-dump.html
8. R. H. Horne and Robin Nagle, 'To Love a Landfill: Dirt and the Environment', in Kate Forde (ed.), *Dirt: The Filthy Reality of Everyday Life* (London: Profile Books, 2009), p. 196.
9. A Greenpeace team identified traces of lead, cadmium, antimony and chlorinated dioxins among other toxins in the charred soil of the Agbogbloshie dump.
10. Tim Brown, *Change by Design: How Design Thinking Transforms Organizations and Inspires Innovation* (New York: HarperCollins, 2009), p. 194.
11. 'Great Pacific Garbage Patch', National Geographic Education, http://education.nationalgeographic.com/education/encyclopedia/great-pacific-garbage-patch/?ar_a=4&ar_r=3
12. William J. Mitchell, Christopher E. Borroni-Bird and Lawrence D. Burns, *Reinventing the Automobile: Personal Mobility for the 21st Century* (Cambridge, Mass.: The MIT Press, 2010), pp. 2–3.
13. Giles Tremlett, 'Madrid Reverses the Chargers with Electric Car Plan', *Guardian*, 8 September 2009, http://www.guardian.co.uk/environment/2009/sep/08/electric-car-plan-spain
14. Sally McGrane, 'Copenhagen Journal: Commuters Pedal to Work on Their Very Own Superhighway', *The New York Times*, 17 July 2012, http://www.nytimes.com/2012/07/18/world/europe/in-denmark-pedaling-to-work-on-a-superhighway.html
15. 'What Goes Around', *Economist*, 19 November 2011, http://www.economist.com/node/21543220
16. Mitchell, Borroni-Bird and Burns, *Reinventing the Automobile*, pp. 38–51.
17. 'Morals and the Machine', *Economist*, 2 June 2012, http://www.economist.com/node/21556234; 'Driverless Cars and How They Would Change Motoring', BBC News, 10 May 2012, http://www.bbc.co.uk/news/magazine-18012812
18. Zipcar, http://www.zipcar.com; Vélib', http://en.velib.paris.fr/; Ecobici, http://www.ecobici.df.gob.mx/; 'Shifting up a Gear', *Economist*, 15 July 2010, http://www.economist.com/node/16591116
19. Raymond Loewy, *Industrial Design* (London: Faber and Faber, 1979), p. 8.
20. Henry Dreyfuss, *Designing for People* (New York: Simon & Schuster, 1955).
21. K. Michael Hays, 'Fuller's Geological Engagements with Architecture', in K. Michael Hays and Dana Miller (eds), *Buckminster Fuller: Starting with the Universe* (New York: Whitney Museum of American Art, 2008), pp. 2–3.
22. Charlotte and Peter Fiell, *1000 Chairs* (Cologne: Taschen, 2000), p. 222.
23. *Ibid.*, pp. 153 and 216.
24. Aldo Leopold, *A Sand County Almanac: And Sketches Here and There* (1949;

Oxford: Oxford University Press, 1992); Rachel Carson, *Silent Spring* (1962; London: Penguin Classics, 2000).

25. Glenn Adamson, 'J.B. Blunk: California Spirit', *Woodwork* (October 1999).
26. Alastair Fuad-Luke, *Design Activism: Beautiful Strangeness for a Sustainable World* (London: Earthscan, 2009), pp. 45–6.
27. Victor Papanek, *Design for the Real World: Human Ecology and Social Change* (1971; Chicago, Ill.: Academy Chicago Publishers, 1985), pp. ix and xi–xvi.
28. Bruce Mau Design, http://www.brucemaudesign.com; 'Work with John Thackara', http://www.doorsofperception.com/working-with-john-thackara/; DESIS Network, http://www.desis-network.org/
29. INDEX: Design to Improve Life, http://www.designtoimprovelife.dk/; The Buckminster Fuller Challenge, http://challenge.bfi.org/
30. 'About Us', Daily Dump, http://www.dailydump.org/about
31. 'Poonam Bir Kasturi: Designing the Daily Dump', Eco Walk the Talk, http://www.ecowalkthetalk.com/blog/2010/07/25/poonam-bir-kasturi-designing-the-daily-dump/
32. 'FAQs', Daily Dump, http://www.dailydump.org/faqs
33. 'Services', Daily Dump, http://www.dailydump.org/services
34. 'Clone Daily Dump', Daily Dump, http://www.dailydump.org/clone_daily_dump
35. 'Home', FARM:, http://farmlondon.weebly.com/
36. Something & Son, http://www.somethingandson.com/
37. Interview with Andy Merritt and Paul Smyth, September 2011, in Alice Rawsthorn, 'Making Food Seriously Local', *International Herald Tribune*, 18 September 2011, http://www.nytimes.com/2011/09/19/arts/19iht-DESIGN19.html
38. Interface, http://www.interface.com
39. 'The Carpet-tile Philosopher', *Economist*, 10 September 2011, http://www.economist.com/node/21528583
40. Paul Hawken *The Ecology of Commerce: A Declaration of Sustainability* (New York: HarperBusiness, 2002).
41. Paul Vitello, 'Ray Anderson, Businessman Turned Environmentalist, Dies at 77', *The New York Times*, 10 August 2011, http://www.nytimes.com/2011/08/11/business/ray-anderson-a-carpet-innovator-dies-at-77.html
42. Cornelia Dean, 'Ray Anderson: Executive on a Mission: Saving the Planet', *The New York Times*, 22 May 2007, http://www.nytimes.com/2007/05/22/science/earth/22ander.html
43. Ray Anderson used this phrase when addressing a group of business people in Toronto in 2005. Vitello, 'Ray Anderson, Businessman Turned Environmentalist, Dies at 77'.
44. Ray Anderson, *Confessions of a Radical Industrialist: How Interface Proved That You Can Build a Successful Business Without Destroying the Planet* (New York: Random House, 2010).
45. 'The Carpet-tile Philosopher', *Economist*, 10 September 2011, http://www.economist.com/node/21528583

11 Why form no longer follows function

1. Paola Antonelli (ed.), *Talk to Me: Design and Communication between People and Objects* (New York: Museum of Modern Art, 2011), pp. 7–8.
2. The phrase 'form follows function' has also been misattributed to the nineteenth-century American sculptor Horatio Greenough. The thinking behind the phrase is reflected in Greenough's writing on art, design and architecture, but he did not use those specific words. However, another source of confusion is that a collection of Greenough's essays published in 1947 is entitled *Form and Function*: Harold A. Small (ed.), *Form and Function: Remarks on Art, Design and Architecture by Horatio Greenough* (Berkeley: University of California Press, 1947).
3. Louis H. Sullivan, 'The Tall Office Building Artistically Considered', *Lippincott's Magazine* (March 1896).
4. The original title of Charles Darwin's book when it was first published in 1859 was *On the Origin of Species by Means of Natural Selection, or the Preservation of Favoured Races in the Struggle for Life*. The short title of the book was abbreviated to *The Origin of Species* upon the publication of the sixth edition in 1872.

5. Henk Tennekes, *The Simple Science of Flight: From Insects to Jumbo Jets* (Cambridge, Mass.: The MIT Press, 2009).
6. The phrase 'less is more' is usually attributed to Mies van der Rohe, who did use it, but it previously appeared in an 1855 poem, 'Andrea del Sarto', by Robert Browning. The poem is written as a dramatic monologue in which Del Sarto, an Italian artist during the Renaissance, addresses his wife Lucrezia del Fede with: 'Well, less is more, Lucrezia: I am judged.' 'Andrea del Sarto', Robert Browning, The Poetry Foundation, http://www.poetryfoundation.org/poem/173001
7. 'Axes', Fiskars UK, http://eng-uk.fiskars.com/Products/Wood-Preparation/Axes
8. Michael Sheridan, *Room 606: The SAS House and the Work of Arne Jacobsen* (London: Phaidon, 2003), pp. 246–9.
9. When it came to improving the performance of a soup spoon, Arne Jacobsen knew exactly what was needed. He loved soup, and consumed it for lunch most days. Usually, he took a portion of soup to his design studio in one of the stainless-steel cocktail shakers he had designed for the Danish company Stelton.
10. Sheridan, *Room 606*, p. 247.
11. Henry Dreyfuss, *Designing for People* (New York: Simon & Schuster, 1955), p. 71.
12. Dieter Rams wrote the 'Ten Principles of Design' during the 1980s at the suggestion of Braun's then head of communications, who recommended that he define the design values of the company for the benefit of the members of its board of directors and the design team, and Rams's students. Originally Rams chose 'The Ten Commandments on Design' as the English name, but came to regard this as too preacherly, and changed it to the 'Ten Principles of Good Design' in 2003. The wording of each point was refined too, but the meaning remained consistent. Dieter Rams, *Less but Better* (Hamburg: Jo Klatt Design+Design, 1995), p. 7.
13. 'Moore's law' is named after Gordon E. Moore, a co-founder of Intel, who published a paper in 1965 noting that the number of components squeezed into integrated circuits had doubled every year since the microchip's invention in 1958. He predicted that this trend would continue for at least another ten years. Moore was correct, and his 'law' has remained accurate ever since. However, current estimates expect the rate of increase to start to slow.
14. James Gleick, *The Information: A History, a Theory, a Flood* (London: Fourth Estate, 2012), pp. 88–124.
15. George Dyson, *Turing's Cathedral: The Origins of the Digital Universe* (London: Allen Lane, 2012), p. 243.
16. The first stored program computer to be developed at the University of Manchester was the Small-Scale Experimental Machine, nicknamed 'Baby', which completed its first program in June 1948. Alan Turing, who had been working on a similar machine at the National Physics Laboratory in Teddington, Middlesex, moved to Manchester and contributed to the development of the next machine, the Manchester Mark 1.
17. Dyson, *Turing's Cathedral*, pp. 257–60. A persistent myth among Apple devotees is that the company's choice of name is a reference to Alan Turing's tragic death in 1954. Turing committed suicide after being convicted of 'gross indecency' with another man and subjected to chemical castration. A half-eaten apple was found near his corpse, fuelling suspicions that he had laced it with cyanide to take a fatal dose of poison.
18. Dyson, *Turing's Cathedral*, pp. ix–x.
19. 'IBM 701', IBM Archives, http://www-03.ibm.com/ibm/history/exhibits/701/701_intro.html
20. 'IBM 5100 Portable Computer', IBM Archives, http://www-03.ibm.com/ibm/history/exhibits/pc/pc_2.html; http://www-03.ibm.com/ibm/history/exhibits/pc/pc_3.html
21. Smaller computers to be used by one person had existed since the 1950s, when a former insurance analyst, Edmund C. Berkeley, developed one named Simon and set up in business as Berkeley Associates in Massachusetts to sell it as a kit of parts for $500 each. Simon could only do simple sums, but its successor, the Geniac Electric Brain Construction Kit, which was introduced by Berkeley in 1955 for $17.95, solved puzzles and played games too. But the 'personal computer' remained a hobbyists' cult until the turn of the 1980s. Mark Frauenfelder, *The Computer* (London: Carlton Books, 2005), pp. 114–43.

22. The mouse was developed in the 1960s by an American technologist, Doug Engelbart, with Bill English, a colleague at the Stanford Research Institute in Menlo Park, California. The name was coined because the shape of their tiny wheeled pointing device resembled the body of a real mouse and its long skinny cord looked like a mouse's tail. Bill Moggridge, *Designing Interactions* (Cambridge, Mass.: The MIT Press, 2007), pp. 17–18.
23. *Ibid.*, pp. 53–4.
24. Walter Isaacson, *Steve Jobs* (London: Little, Brown, 2011), p. 127.
25. Moggridge, *Designing Interactions*, p. 101.
26. The '1984' commercial for the Apple Macintosh was broadcast during the Superbowl on 22 January 1984. '"1984" Apple Macintosh Commercial (Full advert, Hi-Quality)', uploaded on to YouTube by miniroll32 on 27 August 2008, http://www.youtube.com/watch?v=HhsWzJo2sN4
27. 'Minority Report (2002)', http://www.imdb.com/title/tt0181689/
28. Rob Walker, 'Freaks, Geeks and Microsoft: How Kinect Spawned a Commercial Ecosystem', *The New York Times Magazine*, 31 May 2012, http://www.nytimes.com/2012/06/03/magazine/how-kinect-spawned-a-commercial-ecosystem.html
29. Among the first personal monitoring devices in the form of sensor-controlled wristbands were Nike's Fuelband and Jawbone's UP.
30. David Rothenberg, *Survival of the Beautiful: Art, Science, and Evolution* (New York: Bloomsbury, 2011), pp. 5–6.
31. Charles Darwin defined his theory of 'sexual selection' in his 1871 book *The Descent of Man: Selection in Relation to Sex* by explaining how the females of some species are instinctively attracted to particular traits in the males that seemingly serve no practical function, as peahens are to the spectacular tails of peacocks. The value of the male's highly aestheticized appearance lies in its appeal to the female, and thereby in its role in perpetuating the species through breeding. Charles Darwin, *The Descent of Man: Selection in Relation to Sex* (1871; London: Penguin Classics, 2004).
32. Kraftwerk's single 'Pocket Calculator', co-written by Karl Bartos, Ralf Hütter and Emil Schult, was released in 1981 in seven different languages. It appeared on the band's 1981 album *Computer World*. 'Kraftwerk – Pocket Calculator', uploaded on to YouTube by scatmanjohn3001 on 2 August 2009, http://www.youtube.com/watch?v=eSBybJGZoCU
33. John Maeda, *The Laws of Simplicity: Design, Technology, Business, Life* (Cambridge, Mass.: The MIT Press, 2006), p. 3.
34. 'Apple Unveils the New iPod shuffle', Apple Press Info, http://www.apple.com/pr/library/2006/09/12Apple-Unveils-the-New-iPod-shuffle.html

12 **Me, myself and I**

1. Stefan Zweig, *The Post Office Girl* (1982; London: Sort of Books, 2009), p. 6.
2. The British scientist Tim Berners-Lee unveiled a proposal for a global hypertext project to be known as the World Wide Web in 1989 while working at CERN, the European Particle Physics Laboratory near Geneva. The World Wide Web program was made available to his colleagues at CERN in December 1990, and went live on the Internet in 1991. 'Longer Bio for Tim Berners-Lee', http://www.w3.org/People/Berners-Lee/Longer.html
3. 'Iris Recognition Immigration System (IRIS)', Home Office, http://www.ukba.homeoffice.gov.uk/customs-travel/Enteringtheuk/usingiris/
4. 'Zadie Smith's Rules for Writers', *Guardian*, 22 February 2010, http://www.guardian.co.uk/books/2010/feb/22/zadie-smith-rules-for-writers
5. Jaron Lanier, *You Are Not a Gadget: A Manifesto* (London: Allen Lane, 2010).
6. This phenomenon applies to everyone who uses digital media, but has had a particularly powerful effect on those who were born since the mid 1970s and have grown up with it – the 'Net Generation' or 'Net Geners' as the Canadian business strategist Don Tapscott has dubbed them. Rather than having to adapt their behaviour to accommodate advances in digital technology, 'Net Geners' have never known anything else. Don Tapscott, *Growing Up Digital: The Rise of the Net Generation* (New York: McGraw-Hill Books, 1997).

Research by Don Tapscott suggests that avid gamers also score highly for spatial skills, the ability to manipulate 3D objects and hand–eye coordination. He has argued that this makes them well equipped for careers in design, architecture and medicine, especially in surgery, where gamers have proved adept at the laparoscopic techniques which require them to respond to the images transmitted on to a screen from a tiny camera inside the patient's body. Don Tapscott, *Grown Up Digital: How the Net Generation Is Changing the World* (New York: McGraw-Hill Books, 2009), pp. 101–4.

7. Edward Rothstein, 'Typography Fans Say Ikea Should Stick to Furniture', *The New York Times*, 4 September 2009, http://www.nytimes.com/2009/09/05/arts/design/05ikea.html
8. Dylan Tweney, 'The Undesigned Web', *Atlantic* (November 2011), http://www.theatlantic.com/technology/archive/2010/11/the-undesigned-web/65458/
9. Each stone in a drystone wall is chosen to occupy a particular place in the interlocking structure by dint of its size, shape, weight and texture. Such walls date back to the first farming communities in Greece at the start of the Neolithic Age there in 7,000 BC, when people began to supplement the food they found in the wild by cultivating plants and animals, and are still constructed in the same way today.
   Mariana Cook, *Stone Walls: Personal Boundaries* (Bologna: Damiani Editore, 2011).
10. Frederick Winslow Taylor, *The Principles of Scientific Management* (1911; Mineola, New York: Dover Publications, 2003).
11. Timo de Rijk, *Norm = Form: On Standardisation and Design* (Den Haag: Thieme Art/Foundation Design den Haag, 2010), pp. 48–53.
12. Henry Ford quoted himself as having said 'Any customer can have a car painted any colour that he wants so long as it is black' in his autobiography *My Life and Work* (New York: Doubleday, Page and Company, 1922).
13. 'Martha Reeves & The Vandellas – Nowhere to Run (1965) HD', uploaded on to YouTube by MyMotownTunes0815007 on 13 September 2011, http://www.youtube.com/watch?v=17yfqxoSTFM&feature=fvst
14. Interview with Hella Jongerius, January 2010, in Alice Rawsthorn, 'Daring to Play with a Rich Palette', *International Herald Tribune*, 18 January 2009, http://www.nytimes.com/2010/01/18/arts/18iht-design18.html
15. Hella Jongerius designed Repeat fabric for Maharam in 2002: http://www.jongeriuslab.com/site/html/work/repeat/. She designed the B-Set for the Dutch ceramics manufacturer Royal Tichelaar Makkum in 1997; http://www.jongeriuslab.com/site/html/work/b_set/
16. Masaki Kanai, 'Not "This Is What I Want" but "This Will Do"', in Masaki Kanai (ed.), *Muji* (New York: Rizzoli International, 2010), p. 14.
17. Farshid Moussavi, *The Function of Form* (Barcelona: Actar, with Cambridge, Mass.: Harvard University Graduate School of Design, 2009), pp. 18–19.
18. Terence Riley and Barry Bergdoll (eds), *Mies in Berlin* (New York: Museum of Modern Art, 2001), pp. 242–3.
19. 'CCTV – Headquarters', OMA, http://oma.eu/news/2012/cctv-completed
20. Peter Cook, 'Plug-In City', in his *Guide to Archigram 1961–1974* (London: Academy Editions, 1994), pp. 110–23.
21. Lucia Allais, 'Interview with Ronan and Erwan Bouroullec', in Ronan Bouroullec and Erwan Bouroullec (eds), *Ronan and Erwan Bouroullec* (London: Phaidon, 2003), pp. 47–8. An equally versatile modular product developed by Ronan and Erwan Bouroullec is Algue, a spindly twig of plastic, which is ten inches long and just over twelve inches wide. It was designed by them in 2004 and manufactured by Vitra. Any number of pieces can be slotted together to create screens, friezes or room dividers of different shapes, sizes and densities. http://www.bouroullec.com
22. The Swedish designer Reed Kram and his German colleague Clemens Weisshaar began the Breeding Tables project in 2003: http://www.kramweisshaar.com/projects/breeding-tables.html
23. Reed Kram and Clemens Weisshaar designed the My Private Sky project for the German porcelain maker Nymphenburg in 2007: http://www.kramweisshaar.com/projects/my-private-sky.html
24. Tomáš Gabzdil Libertíny developed the first version of the Honeycomb Vase as his

graduation project at Design Academy Eindhoven in the Netherlands in 2006: http://www.tomaslibertiny.com/?portfolio=the-honeycomb-vase
25. Nike Store, http://www.nike.com/us/en_us/lp/nikeid
26. Local Motors, http://www.local-motors.com/
27. 'Print Me a Stradivarius', *Economist*, 12 February 2011, http://www.economist.com/node/18114327
28. A group of designers and software developers at Digital Forming in London deploys 3D printing to enable people to personalize simple plastic objects including bowls, vases and pens. They can only be adapted to a limited degree, because the designers have specified how far the form of each one can be changed before it becomes useless or unstable. http://www.digitalforming.com/index.html
29. 'Special Reports: The Third Industrial Revolution', *Economist*, 21 April 2012, http://www.economist.com/node/21552901
30. Moussavi, *The Function of Form*, p. 25.
31. The Belgian designer Thomas Lommée formed the OpenStructures network in Brussels to develop collaborative design and production systems; http://www.openstructures.net/. Unfold, which is based in Antwerp, was founded in 2002 by Claire Warnier and Dries Verbruggen after graduating from Design Academy Eindhoven; http://unfold.be/pages/projects

13 What about 'the other 90%'?

1. Interview with Paul Polak, April 2007, in Alice Rawsthorn, 'Design for the Unwealthiest 90 Per Cent', *International Herald Tribune*, 30 April 2007, http://www.nytimes.com/2007/04/27/style/27iht-design30.1.5470390.html
2. Interview with Emily Pilloton, October 2011, in Alice Rawsthorn, 'Humanitarian Design Project Aims to Build a Sense of Community', *International Herald Tribune*, 24 October 2011, http://www.nytimes.com/2011/10/24/arts/24iht-design24.html
3. Project H Design, http://www.projecthdesign.org/#studio-h
4. Interview with Emily Pilloton, August 2009, in Alice Rawsthorn, 'Design for Humanity' *International Herald Tribune*, 7 September 2009. http://www.nytimes.com/2009/09/07/fashion/07iht-design7.html
5. 'About Architecture for Humanity', http://architectureforhumanity.org/about
6. Interview with Emily Pilloton, August 2009, in Rawsthorn, 'Design for Humanity'.
7. 'Kutamba AIDS Orphans School', http://openarchitecturenetwork.org/node/1756
8. Emily Pilloton, *Design Revolution: 100 Products That Empower People* (New York: Metropolis Books, 2009), pp. 160–61.
9. Interview with Emily Pilloton, August 2009, in Rawsthorn, 'Design for Humanity'.
10. Interview with Emily Pilloton, August 2010, in Alice Rawsthorn, 'Putting New Tools in Students' Hands', *International Herald Tribune*, 23 August 2010, http://www.nytimes.com/2010/08/23/arts/23iht-design23.html
11. Interview with Emily Pilloton, January 2012. To promote Pilloton's book on role models of humanitarian and sustainable design, *Design Revolution*, she and Miller converted the Airstream into a mobile design gallery, which they drove around the United States with Junebug, a border collie they bought on impulse in Harper, Texas. http://www.projecthdesign.org/#design-revolution-road-show; 'Need to Know | Design Revolution Road Show | PBS', uploaded on to YouTube by PBS on 13 May 2010, http://www.youtube.com/watch?v=lGdRHykBY8A
12. Interview with Emily Pilloton, October 2011, in Rawsthorn, 'Humanitarian Design Project Aims to Build a Sense of Community'.
13. *Ibid.*
14. 'Recap of Grand Opening Ceremony', Studio H, http://www.studio-h.org/recap-of-grand-opening-ceremony
15. Interview with Emily Pilloton, October 2011, in Rawsthorn, 'Humanitarian Design Project Aims to Build a Sense of Community'.
16. Interview with Emily Pilloton, January 2012.
17. Interview with Emily Pilloton, October 2011, in Rawsthorn, 'Humanitarian Design Project Aims to Build a Sense of Community'.

18. The principal funders for year one of Studio H in Bertie County were the W. K. Kellogg Foundation and the Adobe Foundation. 'From Bertie to Berkeley: The Next Generation of Studio H', Studio H, http://www.studio-h.org/from-bertie-to-berkeley-the-next-generation-of-studio-h
19. Paul Polak, founder of International Design Enterprises, coined the term 'design for the other 90%' when speaking at the Aspen Design Summit 2006. 'India', http://blog.paulpolak.com/?cat=11
20. Rick Poynor (ed.), *Communicate: Independent British Graphic Design since the Sixties* (London: Barbican Art Gallery and Laurence King, 2004), pp. 22–3.
21. The many other examples of strategic social design projects include We Are What We Do's work in developing a Young Activist Programme to raise political awareness among young people in Britain. ReD Associates has worked with Copenhagen City council to reduce absenteeism by its workforce, and with the Danish government to try to foster greater understanding of the West among people in the Muslim world.
22. Interview with Hilary Cottam, April 2012.
23. Hilary Cottam set up Participle in partnership with Hugo Manassei, an Internet entrepreneur who is now co-principal partner with her. 'Hilary Cottam', Participle, http://www.participle.net/about/people/24/hilaryc/
24. Interview with Hilary Cottam, April 2012.
25. Interview with Hilary Cottam, October 2008 in Alice Rawsthorn, 'A New Design Concept: Creating Social Solutions for Old Age', *International Herald Tribune*, 27 October 2008, http://www.nytimes.com/2008/10/27/arts/27iht-design27.1.17228735.html
26. Interview with Hilary Cottam, April 2012.
27. Interview with Hilary Cottam, October 2008, in Rawsthorn, 'A New Design Concept'.
28. 'The Circle Movement', Participle, http://www.participle.net/projects/view/5/101/
29. Eric Kindel and Sue Walker with Christopher Burke, Matthew Eve and Emma Minns, 'Isotype Revisited', originally published in Italian translation in *Progetto grafico*, 18 (September 2010); http://www.isotyperevisited.org/2010/09/isotype-revisited.html
30. 'Visual Education Expert Visits Ibadan Schools', an article by an unknown author on Marie Reidemeister's work in West Africa, originally submitted to Nigerian newspapers for publication in 1954, posted on Isotype Revisited, http://www.isotyperevisited.org/1954/07/visual-education-expert-visits-ibadan-schools.html
31. Amelia Gentleman, 'Letter from India: Avant-garde City of Chandigarh, India, Loses Overlooked Treasure', *The New York Times*, 29 February 2008, http://www.nytimes.com/2008/02/29/world/asia/29iht-letter.1.10571360.html
32. Jason Burke, 'Le Corbusier's Indian masterpiece Chandigarh Is Stripped for Parts', *Guardian*, 7 March 2011, http://www.guardian.co.uk/artanddesign/2011/mar/07/chandigarh-le-corbusier-heritage-site
33. Paul Polak, *Out of Poverty: What Works When Traditional Approaches Fail* (San Francisco, Calif.: Berrett-Koehler, 2008).
34. 'History of Architecture for Humanity', http://architectureforhumanity.org/about/history
35. Dan Rockhill and Jenny Kivett, 'Studio 804 in Greensburg, Kansas', in Marie J. Aquilino (ed.), *Beyond Shelter: Architecture and Human Dignity* (New York: Metropolis Books, 2011), pp. 234–45; Studio 804, http://studio804.com/
36. Acumen Fund, http://www.acumenfund.org/
37. John Thackara, *In the Bubble: Designing in a Complex World* (Cambridge, Mass.: The MIT Press, 2006); Alex Steffen (ed.), *Worldchanging: A User's Guide for the 21st Century* (New York: Harry N. Abrams, 2006); Kate Stohr and Cameron Sinclair (eds), *Design Like You Give a Damn: Architectural Responses to Humanitarian Crises* (New York: Metropolis Books, 2006).
38. 'India', http://blog.paulpolak.com/?cat=11
39. Exhibitions Archive, Smithsonian Cooper-Hewitt, National Design Museum in New York, http://archive.cooperhewitt.org/other90/other90.cooperhewitt.org/
40. 'Nicholas Negroponte', MIT Media Lab, http://www.media.mit.edu/people/nicholas; John Markoff, 'New Economy: Taking the Pulse of Technology at Davos', *The New York Times*, 31 January 2005, http://www.nytimes.com/2005/01/31/technology/31newcon.html

41. One Laptop per Child (OLPC), http://laptop.org/en/vision/project/index.shtml
42. John Markoff, 'Microsoft Would Put Poor Online by Cellphone', *The New York Times*, 30 January 2006, http://www.nytimes.com/2006/01/30/technology/30gates.html
43. Interview with Nicholas Negroponte, November 2006, in Alice Rawsthorn, 'One Laptop Per Child: Computer Designed for Those Who Can Least Afford Them', *International Herald Tribune*, 20 November 2006, http://www.nytimes.com/2006/11/19/style/19iht-design20.html
44. One Laptop Per Child (OLPC), http://laptop.org/en/vision/project/index.shtml
45. Interview with Nicholas Negroponte, November 2006, in Rawsthorn, 'One Laptop Per Child'.
46. Interview with Yves Béhar, November 2006, in Rawsthorn, 'One Laptop Per Child'.
47. Interview with Nicholas Negroponte, May 2008, in Alice Rawsthorn, 'Design Accolades for One Laptop Per Child', *International Herald Tribune*, 19 May 2008, http://www.nytimes.com/2008/05/16/arts/16iht-design19.1.12963222.html
48. One Laptop per Child (OLPC), http://laptop.org/en/vision/project/index.shtml
49. Interview with Yves Béhar, May 2008, in Rawsthorn, 'Design Accolades for One Laptop Per Child'.
50. Interview with Nicholas Negroponte, November 2009, in Alice Rawsthorn, 'Nonprofit Laptops: A Dream Not Yet Over', *International Herald Tribune*, 9 November 2009, http://www.nytimes.com/2009/11/09/arts/09iht-design9.html
51. 'Education in Uruguay: Laptops for All', *Economist*, 3 October 2009, http://www.economist.com/node/14558609
52. Interview with Suneet Singh Tuli, chief executive of Datawind, December 2011, in Alice Rawsthorn, 'A Few Stumbles on the Road to Connectivity', *International Herald Tribune*, 19 December 2011, http://www.nytimes.com/2011/12/19/arts/design/a-few-stumbles-on-the-road-to-connectivity.html
53. 'PlayPumps International', uploaded on to YouTube by National Geographic on 9 January 2008, http://www.youtube.com/watch?v=qjgcHOWcWGE
54. Ralph Borland, 'The Problem with the PlayPump', in Ralph Borland, Michael John Gorman, Bruce Misstear and Jane Withers, *Surface Tension: The Future of Water* (Dublin: Science Gallery, 2011), pp. 70–71.
55. Andrew Chambers, 'Africa's Not-so Magic Roundabout', *Guardian*, 24 November 2009, http://www.guardian.co.uk/commentisfree/2009/nov/24/africa-charity-water-pumps-roundabouts. After PlayPumps International closed, its inventory was given to the non-profit group Water for People, based in Denver, Colorado, which is committed to supporting the development of locally sustainable drinking-water resources and sanitation facilities for people in developing countries; http://www.waterforpeople.org/extras/playpumps/update-on-playpumps.html
56. Bruce Nussbaum, 'Is Humanitarian Design the New Imperialism? Does Our Desire to Help Do More Harm than Good?', *Co.Design*, 7 July 2010, http://www.fastcodesign.com/1661859/is-humanitarian-design-the-new-imperialism
57. Nathaniel Corum studied product design for his first degree at Stanford University, took a master's in architecture at the University of Texas, and then worked with tribal groups, first with Berber communities in Morocco and then, back in the US, in Montana and North Dakota. In Montana he met Cameron Sinclair and Kate Stohr, and joined them at Architecture for Humanity when they opened an office in the Bay Area in 2006. He has since developed AfH's education programme, and continued his work with tribal groups, undertaking a long-term project to build sustainable, off-grid homes for Navajo elders in New Mexico and Arizona. Describing himself as 'nomadic', Corum has a base near AfH's office, but spends most of the year travelling from project to project, whether it is contributing to AfH's reconstruction work after the Haiti earthquake; setting up education programmes at schools and universities in Australia and New Zealand; or participating in ad hoc ventures like *The Plastiki*'s voyage. If he has any spare time, he usually goes to New Mexico and Arizona to work on the elders' homes. Interview with Nathaniel Corum, July 2010, in Alice Rawsthorn, 'A Font of Ideas from a "Nomadic" Humanitarian Architect', *International Herald Tribune*, 2 August 2010, http://www.nytimes.com/2010/08/02/arts/design/02iht-design2.html

58. Alongside its work for OLPC, fuseproject embarked upon a safe-sex programme in New York, and developed a free glasses scheme for children in Mexico. Interview with Yves Béhar, December 2011, in Rawsthorn, 'A Few Stumbles on the Road to Connectivity'.
59. Victor Papanek, *Design for the Real World: Human Ecology and Social Change* (1971; Chicago, Ill.: Academy Chicago Publishers, 1985), p. xvii.
60. Interview with Sanga Moses, July 2012.
61. 'About Us', Eco-Fuel Africa, http://www.ecofuelafrica.com/index.php?option=com_content&view=article&id=59&Itemid=882
62. 'Eco-Fuel Africa Limited', The Buckminster Fuller Challenge, http://challenge.bfi.org/2012Finalist_EcoFuel
63. 'Our Team', Eco-Fuel Africa, http://www.ecofuelafrica.com/index.php?option=com_content&view=article&id=62:our-team&catid=21:about-ecofuel&Itemid=887
64. Interview with Sanga Moses, July 2012.

Epilogue **Redesigning design**

1. Emily Pilloton, *Design Revolution: 100 Products That Empower People* (New York: Metropolis Books, 2009), p. 10.
2. Interview with Mathieu Lehanneur, January 2012, in Alice Rawsthorn, 'Blending Fields, Connecting Ideas', *International Herald Tribune*, 16 January 2012, http://www.nytimes.com/2012/01/16/arts/16iht-design16.html
3. Coralie Gauthier (ed.), *Mathieu Lehanneur* (Berlin: Gestalten, 2012), pp. 57–65.
4. Interview with Mathieu Lehanneur, July 2012.
5. 'Design Takes Over, Says Paola Antonelli', *Economist*, 22 November 2010, http://www.economist.com/node/17509367
6. William Little, H. W. Fowler and Jessie Coulson, *The Shorter Oxford English Dictionary on Historical Principles*, vol. 1, ed. C. T. Onions (Oxford: Clarendon Press, 1987), p. 528.

Bibliography

- Stanley Abercrombie, *George Nelson: The Design of Modern Design* (1995; Cambridge, Mass.: The MIT Press, 2000)
- Glenn Adamson and Jane Pavitt (eds), *Postmodernism: Style and Subversion, 1970–1990* (London: V&A Publishing, 2011)
- Ray Anderson, *Confessions of a Radical Industrialist: How Interface Proved That You Can Build a Successful Business Without Destroying the Planet* (New York: Random House, 2010)
- Ed Annink and Max Bruinsma (eds), *Gerd Arntz: Graphic Designer* (Rotterdam: 010 Publishers, 2010)
- Paola Antonelli, *Humble Masterpieces: 100 Everyday Marvels of Design* (2005; London: Thames & Hudson, 2006)
- Paola Antonelli (ed.), *Design and the Elastic Mind* (New York: Museum of Modern Art, 2008)
- Paola Antonelli (ed.), *Safe: Design Takes on Risk* (New York: Museum of Modern Art, 2006)
- Paola Antonelli (ed.), *Talk to Me: Design and Communication between People and Objects* (New York: Museum of Modern Art, 2011)
- Marie J. Aquilino (ed.), *Beyond Shelter: Architecture and Human Dignity* (New York: Metropolis Books, 2011)
- Phil Baines, *Penguin by Design: A Cover Story, 1935–2005* (London: Allen Lane, 2005)
- Reyner Banham, *Design by Choice*, Penny Sparke (ed.) (London: Academy Editions, 1981)
- Reyner Banham, *Theory and Design in the First Machine Age* (1960; Oxford: Butterworth Architecture, 1992)
- Stephen Banham, *Characters: Cultural Stories Revealed through Typography* (Melbourne: Thames & Hudson, 2011)
- Roland Barthes, *The Fashion System* (1967; Berkeley: University of California Press, 1990)
- Roland Barthes, *Mythologies* (1957; Frogmore, St Albans: Paladin, 1973)
- Jennifer Bass and Pat Kirkham, *Saul Bass: A Life in Film & Design* (London: Laurence King, 2011)
- Jean Baudrillard, *The System of Objects* (1968; London: Verso, 2005)
- Bauhaus-Archiv Berlin, Stiftung Bauhaus Dessau and Klassik Stiftung Weimar, *Bauhaus: A Conceptual Model* (Ostfildern: Hatje Cantz, 2009)
- Herbert Bayer, Ise Gropius and Walter Gropius (eds), *Bauhaus 1919–1928* (New York: Museum of Modern Art, 1938)
- Honor Beddard and Douglas Dodds, *Digital Pioneers* (London: V&A Publishing, 2009)
- Marshall Berman, *All That Is Solid Melts into Air: The Experience of Modernity* (London: Verso, 1990)
- Anthony Bertram, *Design* (Harmondsworth: Penguin Books, 1938)
- Regina Lee Blaszczyk, *The Color Revolution* (Cambridge, Mass.: The MIT Press, 2012)
- Andrew Blauvelt and Ellen Lupton (eds), *Graphic Design: Now in Production* (Minneapolis, Minn.: Walker Art Center, 2011)
- Florian Böhm (ed.), *KGID Konstantin Grcic Industrial Design* (London: Phaidon, 2005)
- Olivier Boissière, *Starck®* (Cologne: Benedikt Taschen, 1991)
- Irma Boom (ed.), *Irma Boom: Biography in Books* (Amsterdam: University of Amsterdam Press, 2010)
- Charles Booth, *Life and Labour of the People in London: Volume 1* (London: Macmillan, 1902)
- Achim Borchardt-Hume (ed.), *Albers and Moholy-Nagy: From the Bauhaus to the New World* (London: Tate Publishing, 2006)
- Ralph Borland, Michael John Gorman, Bruce Misstear and Jane Withers, *Surface Tension: The Future of Water* (Dublin: Science Gallery, 2011)

- Ronan Bouroullec and Erwan Bouroullec (eds), *Ronan and Erwan Bouroullec* (London: Phaidon, 2003)
- Nicolas Bourriaud, *Relational Aesthetics* (Dijon: Les Presses du Réel, 1998)
- Charles Arthur Boyer and Federica Zanco, *Jasper Morrison* (Paris: Éditions Dis Voir, 1999)
- Michael Braungart, William McDonough, *Cradle to Cradle: Re-making the Way We Make Things* (2002; London: Jonathan Cape, 2008)
- Giovanni Brino, *Carlo Mollino: Architecture as Autobiography* (London: Thames & Hudson, 1987)
- Tim Brown, *Change by Design: How Design Thinking Transforms Organizations and Inspires Innovation* (New York: HarperCollins, 2009)
- Peter Burke, *The Fabrication of Louis XIV* (New Haven, Conn.: Yale University Press, 1992)
- Jason T. Busch and Catherine L. Futter, *Inventing the Modern World: Decorative Arts at the World's Fairs, 1851–1939* (New York: Skira Rizzoli International, 2012)
- Martha Buskirk and Mignon Nixon (eds), *The Duchamp Effect: Essays, Interviews, Round Table* (Cambridge, Mass.: The MIT Press, 1996)
- Mario Carpo, *The Alphabet and the Algorithm* (Cambridge, Mass.: The MIT Press, 2011)
- Rachel Carson, *Silent Spring* (1962; London: Penguin Classics, 2000)
- Germano Celant (ed.), *Espressioni di Gio Ponti* (Milan: Triennale Electa, 2011)
- Edmonde Charles-Roux, *Chanel* (1974; London: Harvill, 1989)
- C. J. Chivers, *The Gun: The AK-47 and the Evolution of War* (London: Allen Lane, 2010)
- Deborah Cohen, *Household Gods: The British and Their Possessions* (New Haven, Conn.: Yale University Press, 2006)
- David A. Cook, *A History of Narrative Film* (New York: W.W. Norton, 1990)
- Mariana Cook, *Stone Walls: Personal Boundaries* (Bologna: Damiani Editore, 2011)
- Peter Cook (ed.), *A Guide to Archigram 1961–1974* (London: Academy Editions, 1994)
- John Cooper, *The Queen's Agent: Francis Walsingham at the Court of Elizabeth I* (London: Faber and Faber, 2011)
- David Crowley and Jane Pavitt (eds), *Cold War Modern: Design 1945–1970* (London: V&A Publishing, 2008)
- Adèle Cygelman, *Palm Springs Modern: Houses in the California Desert* (New York: Rizzoli International, 1999)
- Charles Darwin, *The Descent of Man: Selection in Relation to Sex* (1871; London: Penguin Classics, 2004)
- Charles Darwin, *The Origin of Species* (1859; Ware: Wordsworth Classics, 1998)
- Daniel Defoe, *The King of Pirates* (1719; London: Hesperus Classics, 2002)
- Chris Dercon (ed.), *Carlo Mollino: Maniera Moderna* (Cologne: Walther König, 2011)
- Alexander Dorner, catalogue for Herbert Bayer Exhibition at the London Gallery, 8 April to 1 May 1937
- Henry Dreyfuss, *Designing for People* (New York: Simon & Schuster, 1955)
- Magdalena Droste, Manfred Ludewig and Bauhaus-Archiv (eds), *Marcel Breuer Design* (Cologne: Benedikt Taschen, 1992)
- Anthony Dunne, *Hertzian Tales: Electronic Products, Aesthetic Experience and Critical Design* (London: Royal College of Art Computer-Related Design Research Studio, 1999)
- George Dyson, *Turing's Cathedral: The Origins of the Digital Universe* (London: Allen Lane, 2012)
- George Eliot, *Middlemarch* (1874; London: Penguin Classics, 1985)
- Ignazia Favata, *Joe Colombo and Italian Design of the Sixties* (London: Thames & Hudson, 1988)
- Fulvio and Napoleone Ferrari, *The Furniture of Carlo Mollino* (London: Phaidon, 2006)
- Charlotte and Peter Fiell, *1000 Chairs* (Cologne: Taschen, 2000)
- Beppe Finessi and Cristina Miglio (eds), *Mendini: A Cura Di* (Mantova: Maurizio Corraini, 2009)
- Victoria Finlay, *Colour* (London: Sceptre, 2002)

- Alberto Fiz (ed.), *Mendini Alchimie: Dal Controdesign alle Nuove Utopie* (Milan: Mondadori Electa S.p.A., 2010)
- Alan Fletcher, *Picturing and Poeting* (London: Phaidon, 2006)
- Lella Secor Florence, *Our Private Lives: America and Britain* (London: George G. Harrap, 1944)
- Henry Ford and Samuel Crowther, *My Life and Work: An Autobiography of Henry Ford* (New York: Doubleday, Page, 1922)
- Kate Forde (ed.), *Dirt: The Filthy Reality of Everyday Life* (London: Profile Books, 2009)
- Norman Foster (ed.), *Dymaxion Car: Buckminster Fuller* (Madrid: Ivorypress, 2010)
- Celina Fox, *The Arts of Industry in the Age of Enlightenment* (New Haven, Conn.: Yale University Press, 2009)
- Nicholas Fox Weber, *The Bauhaus Group: Six Masters of Modernism* (New York: Alfred A. Knopf, 2009)
- Nicholas Fox Weber and Pandora Tabatabai Asbaghi, *Anni Albers* (New York: Guggenheim Museum Publications, 1999)
- Nicholas Fox Weber and Martin Filler, *Josef + Anni Albers: Designs for Living* (London: Merrell, 2004)
- Mark Frauenfelder, *The Computer* (London: Carlton Books, 2005)
- Christopher Frayling, *Ken Adam and the Art of Production Design* (London: Faber and Faber, 2005)
- Arnd Friedrichs and Kerstin Finger (eds), *The Infamous Chair: 220ºC Virus Monobloc* (Berlin: Gestalten, 2010)
- Alastair Fuad-Luke, *Design Activism: Beautiful Strangeness for a Sustainable World* (London: Earthscan, 2009)
- Naoto Fukasawa and Jasper Morrison, *Super Normal: Sensations of the Ordinary* (Baden: Lars Müller, 2007)
- Martino Gamper, *100 Chairs in 100 Days and Its 100 Ways* (London: Dent-De-Leone, 2007)
- Katya García-Antón, Emily King and Christian Brandle, *Wouldn't It Be Nice . . . Wishful Thinking in Art and Design* (Geneva: Centre d'Art Contemporain de Genève, 2007)
- Simon Garfield, *Just My Type: A Book about Fonts* (London: Profile Books, 2010)
- Ken Garland, *Mr Beck's Underground Map* (Harrow: Capital Transport Publishing, 2008)
- Philippe Garner, *Eileen Gray: Designer and Architect* (Cologne: Benedikt Taschen, 1993)
- Elizabeth Gaskell, *North and South* (1855; Harmondsworth: Penguin Classics, 1987)
- Coralie Gauthier (ed.), *Mathieu Lehanneur* (Berlin: Gestalten, 2012)
- Siegfried Giedion, 'The Key to Reality: What Ails Our Time?' Catalogue for Constructivist Art exhibition at the London Gallery, 12–31 July 1937
- Siegfried Giedion, *Space, Time and Architecture: The Growth of a New Tradition* (Cambridge, Mass.: Harvard University Press, 1941)
- Brendan Gill, *Many Masks: A Life of Frank Lloyd-Wright* (London: William Heinemann, 1988)
- James Gleick, *The Information: A History, a Theory, a Flood* (2011; London: Fourth Estate, 2012)
- Andrea Gleiniger, *The Chair No. 14 by Michael Thonet* (Frankfurt am Main, Germany: form, 1998)
- Nicholas Goodison, *Matthew Boulton: Ormolu* (London: Christie's, 1974)
- Robert Graves, *Greek Myths* (1955; London: Cassell, 1991)
- Jean-Pierre Greff (ed.), *AC/DC Contemporary Art, Contemporary Design* (Geneva: Geneva University of Art and Design, 2008)
- Thierry Grillet and Marie-Laure Jousset (eds), *Ettore Sottsass* (Paris: Éditions du Centre Pompidou, 1994)
- Joost Grootens, *I swear I use no art at all: 10 years, 100 books, 18,788 pages of book design* (Rotterdam: 010 Publishers, 2010)
- Robert Grudin, *Design and Truth* (New Haven, Conn.: Yale University Press, 2010)
- Martí Guixé, *Food Designing* (Mantova, Italy: Maurizio Corraini, 2010)
- Martí Guixé (ed.), *Libre de Contexte, Context Free* (Basel: Birkhäuser, 2003)

- Paul Hawken, *The Ecology of Commerce: A Declaration of Sustainability* (New York: HarperBusiness, 2002)
- K. Michael Hays and Dana Miller (eds), *Buckminster Fuller: Starting with the Universe* (New York: Whitney Museum of American Art, 2008)
- Steven Heller, *Paul Rand* (London: Phaidon, 1999)
- John Heskett, *Industrial Design* (London: Thames & Hudson, 1980)
- John Heskett, *Toothpicks & Logos: Design in Everyday Life* (Oxford: Oxford University Press, 2002)
- Fred Hirsch, *The Social Limits to Growth* (London: Routledge & Kegan Paul, 1977)
- E. J. Hobsbawm, *The Age of Capital: 1848–1875* (1975; London: Abacus, 1985)
- E. J. Hobsbawm, *The Age of Extremes: The Short Twentieth Century 1914–1991* (1994; London: Abacus, 2008)
- E. J. Hobsbawm, *The Age of Revolution: Europe 1789–1848* (1962; London: Abacus, 1987)
- E. J. Hobsbawm, *Industry and Empire* (1968; Harmondsworth: Penguin Books, 1982)
- Elaine S. Hochman, *Bauhaus: Crucible of Modernism* (New York: Fromm International, 1997)
- Andrew Hodges, *Alan Turing: The Enigma* (1983; London: Vintage, 2012)
- Anne Hollander, *Sex and Suits: The Evolution of Modern Dress* (1994; Brinkworth: Claridge Press, 1998)
- Richard Holmes, *The Age of Wonder: How the Romantic Generation Discovered the Beauty and Terror of Science* (London: HarperPress, 2008)
- David Hume, *A Treatise of Human Nature: Being an Attempt to Introduce the Experimental Method of Reasoning into Moral Subjects* (1740; London: Penguin Classics, 2004)
- Reginald Isaacs, *Gropius: An Illustrated Biography of the Creator of the Bauhaus* (1983; Boston, Mass.: Bullfinch Press, 1991)
- Walter Isaacson, *Steve Jobs* (London: Little, Brown, 2011)
- Frederic Jameson, *Postmodernism or, The Cultural Logic of Late Capitalism* (London: Verso, 1991)
- Iva Janáková (ed.), *Ladislav Sutnar – Prague – New York – Design in Action* (Prague: Argo, 2003)
- Lisa Jardine, *Worldly Goods: A New History of the Renaissance* (1996; London: Macmillan, 1997)
- Captain Charles Johnson, *A General History of the Robberies and Murders of the Most Notorious Pyrates* (1724; London: Conway Maritime Press, 2002)
- Philip Johnson, *Machine Art* (New York: Museum of Modern Art, 1934)
- Philip Johnson, *Objects: 1900 and Today* (New York: Museum of Modern Art, 1933)
- Steve Jones, *Darwin's Island: The Galapagos in the Garden of England* (London: Little, Brown, 2009)
- Cees W. de Jong (ed.), *Jan Tschichold: Master Typographer, His Life, Work & Legacy* (London: Thames & Hudson, 2008)
- *Donald Judd Furniture* (Rotterdam: Museum Boijmans van Beuningen, 1993)
- Leander Kahney, *Inside Steve's Brain* (New York: Portfolio, 2008)
- Masaki Kanai (ed.), *Muji* (New York: Rizzoli International, 2010)
- Edgar Kaufmann Jnr, *Good Design* (New York: Museum of Modern Art, 1950)
- Edgar Kaufmann Jnr, *Organic Design in Home Furnishings* (New York: Museum of Modern Art, 1941)
- Edgar Kaufmann Jnr, *Prize Designs for Modern Furniture* (New York: Museum of Modern Art, 1950)
- Donald Keene, *Yoshimasa and the Silver Pavilion: The Creation of the Soul of Japan* (New York: Columbia University Press, 2003)
- Alison Kelly (ed.), *The Story of Wedgwood* (1962; London: Faber and Faber, 1975)
- György Kepes, *Language of Vision* (1944; New York: Dover Publications, 1995)
- György Kepes (ed.), *György Kepes: The MIT Years, 1945–1977* (Cambridge, Mass.: The MIT Press, 1978)
- Emily King, *Robert Brownjohn: Sex and Typography* (New York: Princeton Architectural Press, 2005)

- Emily King (ed.), *Designed by Peter Saville* (London: Frieze, 2003)
- Pat Kirkham, *Charles and Ray Eames: Designers of the Twentieth Century* (Cambridge, Mass.: The MIT Press, 1995)
- Naomi Klein, *No Logo* (London: Flamingo, 2000)
- Rem Koolhaas, Bruce Mau with Jennifer Sigler (eds). *Small, Medium, Large, Extra-Large: Office for Metropolitan Architecture* (Rotterdam: 010 Publishers, 1995)
- Rem Koolhaas and Hans Ulrich Obrist, *Project Japan: Metabolism Talks . . .* (Cologne: Taschen, 2011)
- Joachim Krausse and Claude Lichtenstein (eds), *Your Private Sky: R. Buckminster Fuller* (Baden: Lars Müller, 2000)
- Mateo Kries and Alexander von Vegesack (eds), *Joe Colombo: Inventing the Future* (Weil am Rhein: Vitra Design Museum, 2005)
- Peter Lang and William Menking (eds), *Superstudio: Life Without Objects* (Milan: Skira Editore, 2003)
- Jaron Lanier, *You Are Not a Gadget: A Manifesto* (London: Allen Lane, 2010)
- Marcia Lausen, *Design for Democracy: Ballot and Election Design* (Chicago, Ill.: University of Chicago Press, 2007)
- Aldo Leopold, *A Sand County Almanac: And Sketches Here and There* (1949; Oxford: Oxford University Press, 1992)
- Jeremy Lewis, *The Life and Times of Allen Lane* (London: Penguin Books, 2006)
- William Little, H. W. Fowler and Jessie Coulson, *The Shorter Oxford English Dictionary on Historical Principles*, vol. 1, ed. C. T. Onions (Oxford: Clarendon Press, 1987)
- Raymond Loewy, *Industrial Design* (London: Faber and Faber, 1979)
- Raymond Loewy, *Locomotive: The New Vision* (New York: The Studio, 1937)
- Loretta Lorance, *Bocoming Bucky Fuller* (Cambridge, Mass.: The MIT Press, 2009)
- Sophie Lovell, *Dieter Rams: As Little Design as Possible* (London: Phaidon, 2011)
- Jacques Lucan (ed.), *OMA – Rem Koolhaas: Architecture 1970–1990* (New York: Princeton Architectural Press, 1991)
- Fiona MacCarthy, *The Last Pre-Raphaelite: Edward Burne-Jones and the Victorian Imagination* (London: Faber and Faber, 2011)
- Fiona MacCarthy, *William Morris* (1994; London: Faber and Faber, 1995)
- Cara McCarty, *Designs for Independent Living* (New York: Museum of Modern Art, 1988)
- Deborah McDonald, *Clara Collet, 1860–1948: An Educated Working Woman* (London: Routledge, 2004)
- Mary McLeod (ed.), *Charlotte Perriand: An Art of Living* (New York: Harry N. Abrams, 2003)
- John Maeda, *The Laws of Simplicity: Design, Technology, Business, Life* (Cambridge: Mass.: The MIT Press, 2006)
- John Maeda with Becky Bermont, *Redesigning Leadership: Design, Technology, Business, Life* (Cambridge: Mass.: The MIT Press, 2011)
- Karl Mang, *History of Modern Furniture* (Stuttgart: Gerd Hatje, 1978)
- Beate Manske (ed.), *Wilhelm Wagenfeld (1900–1990)* (Ostfildern-Ruit: Hatje Cantz, 2000)
- Abraham H. Maslow (ed.), *New Knowledge in Human Values* (New York: Harper & Brothers, 1959)
- Shena Mason (ed.), *Matthew Boulton: Selling What All the World Desires* (New Haven, Conn.: Yale University Press, 2009)
- Bruce Mau, *Life Style* (London: Phaidon, 2000)
- Bruce Mau and the Institute Without Boundaries, *Massive Change* (London: Phaidon, 2004)
- Christien Meindertsma, *Pig 05049* (Rotterdam: Flocks, 2007)
- Metahaven and Marina Vishmidt, *Uncorporate Identity* (Baden: Lars Müller with the Jan van Eyck Academie, Maastricht, 2010)
- William J. Mitchell, Christopher E. Borroni-Bird and Lawrence D. Burns, *Reinventing the Automobile: Personal Mobility for the 21st Century* (Cambridge, Mass.: The MIT Press, 2010)

- Bill Moggridge, *Designing Interactions* (Cambridge, Mass.: The MIT Press, 2007)
- Bill Moggridge, *Designing Media* (Cambridge, Mass.: The MIT Press, 2010)
- László Moholy-Nagy, *Vision in Motion* (Chicago, Ill.: Paul Theobald, 1947)
- Sibyl Moholy-Nagy, *Moholy-Nagy: Experiment in Totality* (New York: Harper & Brothers, 1950)
- Richard Morphet (ed.), *Richard Hamilton* (London: Tate Gallery Publications, 1992)
- William Morris, *Hopes and Fears for Art* (London: Longmans, Green, 1919)
- Jasper Morrison, *A Book of Spoons* (Gent: Imschoot Uitgevers, 1997)
- Jasper Morrison, *Everything but the Walls* (Baden: Lars Müller, 2002)
- Farshid Moussavi, *The Function of Form* (Barcelona: Actar, with Cambridge, Mass.: Harvard University Graduate School of Design, 2009)
- Farshid Moussavi and Michael Kubo (eds), *The Function of Ornament* (Barcelona: Actar, 2006)
- Bruno Munari, *Design as Art* (1966; London: Penguin Books, 2008)
- Bruno Munari, *Supplemento al Dizionario Italiano/Supplement to the Italian dictionary* (1963; Mantova: Maurizio Corraini, 2004)
- Heike Munder (ed.), *Peter Saville Estate 1–127* (Zurich: Migros Museum für Gegenwartskunst Zürich and JRP|Ringier, 2007)
- George Nelson, *Chairs* (New York: Whitney, 1953)
- John Neuhart, Marilyn Neuhart and Ray Eames, *Eames Design: The Work of the Office of Charles and Ray Eames* (London: Thames & Hudson, 1989)
- Marilyn Neuhart with John Neuhart, *The Story of Eames Furniture* (Berlin: Gestalten, 2010)
- Marie Neurath and Robert S. Cohen (eds), *Otto Neurath: Empiricism and Sociology* (Dordecht: D. Reidel, 1973)
- Otto Neurath, *From Hieroglyphics to Isotype: A Visual Autobiography* (London: Hyphen Press, 2010)
- Jocelyn de Noblet (ed.), *Design, Miroir du Siècle* (Paris: Flammarion/APCI, 1993)
- Novalis, *Novalis: Philosophical Writings* (1798; Albany: State University of New York Press, 1977)
- Hans Ulrich Obrist (ed.), *A Brief History of Curating* (Zurich: JRP|Ringier, 2008)
- Celeste Olalquiaga, *The Artificial Kingdom: A Treasury of the Kitsch Experience* (New York: Pantheon Books, 1998)
- Jonathan Olivares, *A Taxonomy of the Office Chair* (London: Phaidon, 2011)
- Victor Papanek, *Design for the Real World: Human Ecology and Social Change* (1971; Chicago, Ill.: Academy Chicago Publishers, 1985)
- Martin Pawley, *Buckminster Fuller: How Much Does the Building Weigh?* (1990; London: Trefoil Publications, 1995)
- Ingrid Pfeiffer and Max Hollein (eds), *László Moholy-Nagy Retrospective* (Munich: Prestel, 2009)
- Emily Pilloton, *Design Revolution: 100 Products That Empower People* (New York: Metropolis Books, 2009)
- Plato, *Early Socratic Dialogues* (1987; London: Penguin Classics, 2005)
- Plato, *The Republic* (1955; London: Penguin Classics, 2007)
- Paul Polak, *Out of Poverty: What Works When Traditional Approaches Fail* (San Francisco, California: Berrett-Koehler Publishers, 2008)
- Sergio Polano, *Achille Castiglioni: Tutte le Opere 1938–2000* (Milan: Electa, 2001)
- Lisa Licitra Ponti, *Gio Ponti: The Complete Work 1923–1978* (London: Thames and Hudson, 1990)
- Jane Portal (ed.), *The First Emperor: China's Terracotta Army* (London: British Museum Press, 2007)
- Rick Poynor, *No More Rules: Graphic Design and Postmodernism* (London: Laurence King, 2003)
- Rick Poynor (ed.), *Communicate: Independent British Graphic Design since the Sixties* (London: Barbican Art Gallery and Laurence King, 2004)
- Graham Pullin, *Design Meets Disability* (Cambridge, Mass.: The MIT Press, 2009)
- Barbara Radice, *Ettore Sottsass: A Critical Biography* (New York: Rizzoli International, 1993)

- Dieter Rams, *Less but Better* (Hamburg: Jo Klatt Design+Design, 1995)
- Herbert Read, *Art and Industry* (London: Faber and Faber, 1934)
- Herbert Read and Bernard Rackham, *English Pottery* (London: Ernest Benn, 1924)
- Casey Reas and Ben Fry, *Processing: A Programming Handbook for Visual Designers and Artists* (Cambridge, Mass.: The MIT Press, 2007)
- Casey Reas and Chandler McWilliams, LUST, *Form + Code: In Design, Art and Architecture* (New York: Princeton Architectural Press, 2010)
- Robert Recorde, *The Whetstone of Witte* (1557; Mountain View, Calif.: Creative Commons, 2009)
- Peter Reed (ed.), *Alvar Aalto: Between Humanism and Materialism* (New York: Museum of Modern Art, 1998)
- Timo de Rijk, *Norm = Form: On Standardisation and Design* (Den Haag: Thieme Art/ Foundation Design den Haag, 2010)
- Terence Riley and Barry Bergdoll (eds), *Mies in Berlin* (New York: Museum of Modern Art, 2001)
- Marco Romanelli, *Gio Ponti: A World* (Milan: Editrice Abitare Segesta, 2002)
- Jamaica Rose and Michael MacLeod, *A Book of Pirates: A Guide to Plundering, Pillaging and Other Pursuits* (Layton, Utah: Gibbs M. Smith, 2010)
- David Rothenberg, *Survival of the Beautiful: Art, Science, and Evolution* (New York: Bloomsbury, 2011)
- Bernard Rudofsky, *Architecture Without Architects: A Short Introduction to Non-Pedigreed Architecture* (1964; Albuquerque: University of New Mexico Press, 1987)
- Louise Schouwenberg, *Hella Jongerius* (London, New York: Phaidon, 2003)
- Louise Schouwenberg (ed.), *Hella Jongerius: Misfit* (London, New York: Phaidon, 2010)
- Franz Schulze, *Philip Johnson: Life and Work* (New York: Alfred A. Knopf, 1994)
- Sabine Schulze and Ina Grätz (eds), *Apple Design* (Ostfildern: Hatje Cantz, 2011)
- Meryle Secrest, *Frank Lloyd Wright: A Biography* (New York: Alfred A. Knopf, 1992)
- Richard Sennett, *The Conscience of the Eye: The Design and Social Life of Cities* (New York: Alfred A. Knopf, 1990)
- Richard Sennett, *The Craftsman* (London: Allen Lane, 2008)
- Paul Shaw, *Helvetica and the New York City Subway System: The True (Maybe) Story* (Cambridge, Mass.: The MIT Press, 2010)
- Mary Shelley, *Frankenstein: Or, the Modern Prometheus* (1818; London: Penguin Classics, 2003)
- Michael Sheridan, *Room 606: The SAS House and the Work of Arne Jacobsen* (London: Phaidon, 2003)
- Harold A. Small (ed.), *Form and Function: Remarks on Art, Design and Architecture by Horatio Greenough* (Berkeley: University of California Press, 1947)
- Adam Smith, *The Wealth of Nations: Books I–III* (1776; London: Penguin Classics, 2003)
- Félix Solaguren-Beascoa de Corral, *Arne Jacobsen* (Barcelona: Editorial Gustavo Gili, 1991)
- Susan Sontag (ed.), *Barthes: Selected Writings* (London: Fontana, 1983)
- Penny Sparke, *A Century of Car Design* (London: Mitchell Beazley, 2002)
- Penny Sparke, *Italian Design: 1870 to the Present* (London: Thames & Hudson, 1988)
- Nancy Spector (ed.), *Matthew Barney: The Cremaster Cycle* (New York: Harry N. Abrams, 2002)
- Alex Steffen (ed.), *Worldchanging: A User's Guide for the 21st Century* (New York: Harry N. Abrams, 2006)
- Kate Stohr and Cameron Sinclair (eds), *Design Like You Give a Damn: Architectural Responses to Humanitarian Crises* (New York: Metropolis Books, 2006)
- Nina Stritzler-Levine (ed.), *Sheila Hicks: Weaving as Metaphor* (New Haven, Conn.: Yale University Press, 2006)
- Don Tapscott, *Growing Up Digital: The Rise of the Net Generation* (New York: McGraw-Hill Books, 1997)
- Don Tapscott, *Grown Up Digital: How the Net Generation Is Changing the World* (New York: McGraw-Hill Books, 2009)

- Frederick Winslow Taylor, *The Principles of Scientific Management* (1911; New York: Dover Publications, 2003)
- Henk Tennekes, *The Simple Science of Flight: From Insects to Jumbo Jets* (Cambridge, Mass.: The MIT Press, 2009)
- John Thackara, *In the Bubble: Designing in a Complex World* (Cambridge, Mass.: The MIT Press, 2006)
- John Thackara (ed.), *Design After Modernism: Beyond the Object* (London: Thames & Hudson, 1988)
- Ian Thompson, *The Sun King's Garden: Louis XIV, André Le Nôtre and the Creation of the Gardens of Versailles* (London: Bloomsbury, 2006)
- Calvin Tomkins, *Duchamp: A Biography* (New York: Henry Holt, 1996)
- Edward Tufte, *Beautiful Evidence* (Cheshire, Conn.: Graphics Press, 2006)
- Edward Tufte, *Envisioning Information* (Cheshire, Conn.: Graphics Press, 1990)
- Edward Tufte, *The Visual Display of Quantitative Information* (1983; Cheshire, Conn.: Graphics Press, 2001)
- Margarita Tupitsyn (ed.), *Rodchenko & Popova: Defining Constructivism* (London: Tate Publishing, 2009)
- Keiko Ueki-Polet and Klaus Kemp (eds), *Less and More: The Design Ethos of Dieter Rams* (Berlin: Gestalten, 2009)
- Jenny Uglow, *The Lunar Men: The Friends Who Made the Future, 1730–1810* (London: Faber and Faber, 2002)
- Giorgio Vasari, *Lives of the Artists: Volume I* (1550; London: Penguin Books, 1987)
- Alexander von Vegesack, *Thonet: Classic Furniture in Bent Wood and Tubular Steel* (London: Hazar, 1996)
- Robert Venturi, Denise Scott Brown and Steven Izenour, *Learning from Las Vegas: The Forgotten Symbolism of Architectural Form* (1972; Cambridge, Mass.: The MIT Press, 1977)
- Ron van der Vlugt, *Life Histories of 100 Famous Logos* (Amsterdam: BIS Publishers, 2012)
- Alison Weir, *Henry VIII: King and Court* (London: Jonathan Cape, 2001)
- Peter Weiss (ed.), *Alessandro Mendini: Design and Architecture* (Milan: Electa, 2001)
- Nigel Whiteley, *Reyner Banham: Historian of the Immediate Future* (Cambridge, Mass.: The MIT Press, 2002)
- Raymond Williams, *Keywords: A Vocabulary of Culture and Society* (1976; London: Fontana, 1983)
- Edward O. Wilson, *The Diversity of Life* (1992; London: Penguin Books, 2001)
- Elizabeth Wilson, *Adorned in Dreams: Fashion and Modernity* (London: Virago Press, 1985)
- Elizabeth Wilson, *Hallucinations: Life in the Post-Modern City* (London: Hutchinson Radius, 1989)
- Hans M. Wingler, *Bauhaus: Weimar, Dessau, Berlin, Chicago* (Cambridge, Mass.: The MIT Press, 1976)
- Theodore Zeldin, *An Intimate History of Humanity* (1994; London: Vintage, 1998)
- Stefan Zweig, *The Post Office Girl* (1982; London: Sort of Books, 2009)

List of illustrations

Plate 1
Hagar Qim Temple
© 2006 Mariana Cook
The ancient stones at Hagar Qim Temple, Malta
Published in the book *Stone Walls: Personal Boundaries* by Mariana Cook

Limestone Field
© 2005 Mariana Cook
Limestone lace wall on Inis Meáin, Ireland
Published in the book *Stone Walls: Personal Boundaries* by Mariana Cook

My Wall in Snow
© 2007 Mariana Cook
Drystone wall at Mariana Cook's home in Chilmark, Martha's Vineyard
Published in the book *Stone Walls: Personal Boundaries* by Mariana Cook

Stone Wall Detail
© 2003 Mariana Cook
Drystone wall on the Shetland Islands, Scotland
Published in the book *Stone Walls: Personal Boundaries* by Mariana Cook

Plate 2
A working model of the Manchester University Mark I computer in 1948
Courtesy: University of Manchester

IBM 701 mainframe computer
Courtesy: IBM Corporate Archives. Kindly reproduced with permission from IBM

Production of the IBM 701 mainframe computer
Courtesy: IBM Corporate Archives. Kindly reproduced with permission from IBM

IBM System 360 mainframe computer
Courtesy: IBM Corporate Archives. Kindly reproduced with permission from IBM

Apple iPad mini in 2013
Courtesy: Apple

Plate 3
R. Buckminster Fuller with models of the geodesic dome at Black Mountain College, North Carolina, in 1948.
Photography: Hazel Larsen Archer
Courtesy: Black Mountain College Museum + Art Center and the Estate of Hazel Larsen Archer

László Moholy-Nagy in Chicago in 1945
Courtesy: The Estate of László Moholy-Nagy

György Kepes at the Center for Advanced Visual Studies at the Massachusetts Institute of Technology in 1971
Photography: Nishan Bichajian
Courtesy: György Kepes Foundation

Self-portrait of Muriel Cooper taken in the Visible Language Workshop of the Massachusetts Institute of Technology in 1977
Photography: Muriel Cooper

Plate 4
Aimee Mullins wearing silicone legs
Copyright: Jill Greenberg

Matthew Barney
Cremaster 3 (2002)
Production still
© 2002 Matthew Barney
Photography: Chris Winget
Courtesy: Gladstone Gallery, New York and Brussels

Matthew Barney
Cremaster 3 (2002)
Production still
© 2002 Matthew Barney
Photography: Chris Winget
Courtesy: Gladstone Gallery, New York and Brussels

Bespoke wooden legs worn by Aimee Mullins in Alexander McQueen's autumn/winter 1999 women's wear collection fashion show.
Courtesy: Aimee Mullins

Plate 5
Billboard introducing Studio H in Windsor, North Carolina
Courtesy: Project H Design

The high-school auto body shop converted into Studio H's classroom
Courtesy: Project H Design

Matthew Miller teaching a class at Studio H
Courtesy: Project H Design

Emily Pilloton teaching a class at Studio H
Courtesy: Project H Design

Studio H students constructing the Windsor Super Market
Courtesy: Project H Design

The completed Windsor Super Market
Photography: Brad Feinknopf, 2011
Courtesy: Project H Design

Vendors at the Windsor Super Market
Courtesy: Project H Design

Customers at the Windsor Super Market
Photography: Brad Feinknopf, 2011
Courtesy: Project H Design

Plate 6
Thonet's grinding shop in Bystritz, *c.*1920
Courtesy: Thonet Gmbh

A Thonet warehouse in Marseille, *c.*1920
Courtesy: Thonet Gmbh

Transportation for Thonet in Bystritz, *c.*1920
Courtesy: Thonet Gmbh

Plate 7
Customized tricycles in Beijing
Photography: Alice Rawsthorn

Plate 8
Tomáš Gabzdil Libertíny's Honeycomb Vase
Photography: Raoul Kramer
Courtesy: Tomáš Gabzdil Libertiny

Cow benches designed by Julia Lohmann
Courtesy: Julia Lohmann

The *Pig 05049* book designed by Christien Meindertsma and Julie Joliat
Courtesy: Christien Meindertsma

Christoph Büchel
'Simply Botiful' (2006)
Installation views at Hauser & Wirth Coppermill, London, England
Photography: Mike Bruce
Copyright: Christoph Büchel
Courtesy: the artist and Hauser & Wirth

Plate 9
One of the 'Objets Thérapeutiques' designed by Mathieu Lehanneur
Photography: © Véronique Huyghe
Courtesy: Mathieu Lehanneur

An analgesic pen in Mathieu Lehanneur's 'Objets Thérapeutiques' project
Photography: © Véronique Huyghe
Courtesy: Mathieu Lehanneur

Layered antibiotics in Mathieu Lehanneur's 'Objets Thérapeutiques' project
Photography: © Véronique Huyghe
Courtesy: Mathieu Lehanneur

'Tomorrow Is Another Day' designed by Mathieu Lehanneur
Photography: Felipe Ribon
Courtesy: Mathieu Lehanneur

Plate 10
Booth Map 5
Courtesy: The Library of the London School of Economics & Political Science

Modular Isotype exhibition structure by Josef Frank for the Gesellschafts und Wirtschaftsmuseum in Vienna, *c.*1929.
IC 6/2 N1751.
Courtesy: Otto and Marie Neurath Isotype Collection, Department of Typography & Graphic Communication, University of Reading

Marie Reidemeister transforming statistical information into a graphic configuration of marks that will guide the production of an Isotype chart, Vienna, *c.*1930.
IC 6/4 N1708.
Courtesy: Otto and Marie Neurath Isotype Collection, Department of Typography & Graphic Communication, University of Reading

'Humans vs Chimps'
Courtesy: Ben Fry

Data visualization of a day's trading on NASDAQ on 24 July 2012
Courtesy: Stamen Design

Plate 11
FARM:Shop
Courtesy: Something & Son

Plate 12
Stratigraphic Porcelain in production
Photography: Kristof Vrancken
Courtesy: Unfold

Stratigraphic Porcelain
Photography: Kristof Vrancken
Courtesy: Unfold

L'Artisan Électronique
Photography: Kristof Vrancken
Courtesy: Unfold/Z33 House for Contemporary Art

Plate 13
Participle's Circle
Photography: Hannah Maule-ffinch
(www.hannahmauleffinch.com)
Courtesy: Participle Ltd

Plate14
Eco-fuel Africa
Courtesy: Eco-fuel Africa

Index